Practical WebAssembly

Explore the fundamentals of WebAssembly programming using Rust

Sendil Kumar Nellaiyapen

BIRMINGHAM—MUMBAI

Practical WebAssembly

Associate Group Product Manager: Pavan Ramchandani
Publishing Product Manager: Bhavya Rao
Senior Editor: Mark Dsouza
Content Development Editor: Divya Vijayan
Technical Editor: Joseph Aloocaran
Copy Editor: Safis Editing
Project Coordinator: Rashika Ba
Proofreader: Safis Editing
Indexer: Subalakshmi Govindhan
Production Designer: Alishon Mendonca
Marketing Coordinator: Anamika Singh
First published: April 2022

Production reference: 2020522
Published by Packt Publishing Ltd.

Livery Place
35 Livery Street
Birmingham
B3 2PB, UK.
ISBN 978-1-83882-800-4
www.packt.com

To all the amazing developers out there.

Contributors

About the author

Sendil Kumar Nellaiyapen is an engineering manager building web payments at Uber. He is passionate about the web and cloud technologies. He has over 12 years of experience in building distributed, cloud-native, and enterprise systems. He occasionally rambles on his website and speaks at various conferences. He is an active open source contributor and enjoys building communities. He loves to learn and explore new programming languages.

I want to thank the amazing Rust and web community.

About the reviewer

Sufyan bin Uzayr is a writer, teacher, and developer with 10+ years of experience in the industry. He is an open source enthusiast and specializes in a wide variety of technologies. He holds four master's degrees and has authored multiple books. Sufyan is an avid writer. He regularly writes about topics related to coding, tech, politics, and sports. He is a regular columnist for various publications and magazines. Sufyan is the CEO of Parakozm, a software development company catering to a global clientele. He is also the CTO at Samurai Servers, a web server management company focusing mainly on enterprise-scale audiences. In his spare time, Sufyan teaches coding and English to young students. Learn more about his works at `https://sufyanism.com/`.

Table of Contents

Preface

Section 1: Introduction to WebAssembly

1
Understanding LLVM

Technical requirements	4	Exploring LLVM	7
Understanding compilers	4	LLVM in action	8
Compiled languages	5	Summary	10
Compiler efficiency	6		

2
Understanding Emscripten

Technical requirements	13	Listing the tools and SDK	24
Installing Emscripten using emsdk	13	Managing the tools and SDK	25
Generating asm.js using Emscripten	15	Understanding various levels of optimizations	28
Running Hello World with Emscripten in Node.js	17	Optimizations	28
		Closure Compiler	32
Running Hello World with Emscripten in the browser	20	Summary	34
Exploring other options in emsdk	22		

3

Exploring WebAssembly Modules

Technical requirements	36	Exploring the WebAssembly text format	40
Understanding how WebAssembly works	36	Building a function in WebAssembly text format	45
Understanding JavaScript execution inside the JavaScript engine	36	Summary	50
Understanding WebAssembly execution inside the JavaScript engine	38		

Section 2: WebAssembly Tools

4

Understanding WebAssembly Binary Toolkit

Technical requirements	54	Converting WAST into JSON	73
Getting started with WABT	54	Understanding a few other tools provided by WABT	75
Installing WABT	54	wasm-objdump	75
Converting WAST into WASM	58	wasm-strip	78
Converting WASM into WAST	62	wasm-validate	79
-f or --fold-exprs	66	wasm-interp	81
Converting WASM into C	67	Summary	82
simple.h	68		
simple.c	69		

5

Understanding Sections in WebAssembly Modules

Technical requirements	84	Globals	89
Exports and imports	84	Start	95
Exports	84	Memory	96
Imports	87	Summary	100

6
Installing and Using Binaryen

Technical requirements	102	wasm-dis	107
Installing and using Binaryen	102	wasm-opt	109
Linux/macOS	103	wasm2js	111
Windows	103	Summary	113
wasm-as	105		

Section 3: Rust and WebAssembly

7
Integrating Rust with WebAssembly

Technical requirements	118	Installing wasm-bindgen	127
Installing Rust	118	Converting Rust into WebAssembly via wasm-bindgen	129
Converting Rust into WebAssembly via rustc	120		
Converting Rust into WebAssembly via Cargo	123	Summary	135

8
Bundling WebAssembly Using wasm-pack

Technical requirements	138	How to use wasm-pack	150
Bundling WebAssembly modules with webpack	138	Packing and publishing using wasm-pack	153
Bundling WebAssembly modules with Parcel	145	Summary	157
Introducing wasm-pack	150		
Why do you need wasm-pack?	150		

9

Crossing the Boundary between Rust and WebAssembly

Technical requirements	160	Calling closures via WebAssembly	173
Sharing classes from Rust with JavaScript	160	Importing the JavaScript function into Rust	177
Sharing classes from JavaScript with Rust	164	Calling a web API via WebAssembly	180
Calling the JavaScript API via WebAssembly	169	Summary	185

10

Optimizing Rust and WebAssembly

Technical requirements	187	to use in the Rust application	195
Minimizing the WebAssembly modules	188	Analyzing the WebAssembly module with Twiggy	198
Analyzing the memory model in the WebAssembly module	192	top	200
Sharing memory between JavaScript and WebAssembly using Rust	192	monos	202
Creating a memory object in JavaScript		garbage	203
		Summary	204

Index

Other Books You May Enjoy

Preface

Delivering high-performance applications is a nightmare. JavaScript is a dynamically typed language. Thus, the JavaScript engine assumes the type when executing JavaScript. These assumptions lead to unpredictable performance. This makes it even harder to deliver consistently high-performance applications in JavaScript.

WebAssembly provides a way to run type-safe and high-performance applications in the JavaScript engine.

WebAssembly is blazingly fast

WebAssembly is the next great thing that happened in the web. It promises high and consistent performance with maintainable code, running native code and providing near-native performance on the web.

WebAssembly is type-safe

The JavaScript compiler struggles to provide high performance when you have polymorphic JavaScript code. WebAssembly, on the other hand, is type-safe (or monomorphic) at compile time. This not only boosts performance but also greatly reduces runtime errors, which is a win-win.

WebAssembly runs your native code

There have been multiple attempts to make the web faster by running native code. But they all failed because they are either vendor-specific or tied to a single language. The web is built on top of open standards. By being an open standard, WebAssembly makes it easy for all companies to adopt and support it. WebAssembly is not a language; it is a high-level implementation plan for other languages that compile to byte code that will run on the JavaScript engines.

WebAssembly is byte code

WebAssembly is nothing but a bytecode that runs in JavaScript Engine.

In this book, we will learn how to convert native code into WebAssembly and how to optimize it to get even better performance. We will also cover how the entire WebAssembly runs on the JavaScript engine and how to use the various tools available and what they help us to achieve.

Most importantly, learn where and how to use WebAssembly to get the desired result out of it.

Let's make the web even more awesome and faster with WebAssembly.

Who this book is for

This book is for JavaScript developers who want to deliver better performance and ship type-safe code. Rust developers or backend engineers looking to build full-stack applications without worrying too much about JavaScript programming will also find the book useful.

A basic understanding of JavaScript is required to follow along with this book. Rust knowledge is preferred but not mandatory. The code samples are simple and easy for any developer to follow along with.

What this book covers

Chapter 1, Understanding LLVM, gives a brief introduction to LLVM, what it is, and how to use it.

Chapter 2, Understanding Emscripten, introduces you to Emscripten, where you will build and run your first WebAssembly module.

Chapter 3, Exploring WebAssembly Modules, explores the WebAssembly module, what the module consists of, and what the different sections are.

Chapter 4, Understanding WebAssembly Binary Toolkit, explores how to install and use **WebAssembly Binary Toolkit (WABT)**.

Chapter 5, Understanding Sections in WebAssembly Modules, explores various sections inside the WebAssembly binary and what their purpose is.

Chapter 6, Installing and Using Binaryen, explores how to install and use Binaryen.

Chapter 7, Integrating Rust with WebAssembly, starts by looking at Rust and various ways to convert Rust into a WebAssembly module and ends by looking at `wasm_bindgen`.

Chapter 8, Bundling WebAssembly Using wasm-pack, explores `wasm-pack` and how it makes it easy to build Rust and WebAssembly applications.

Chapter 9, Crossing the Boundary between Rust and WebAssembly, focuses on how `wasm-bindgen`, along with crates such as `js-sys` and `web-sys`, helps to share entities from WebAssembly with JavaScript.

Chapter 10, Optimizing Rust and WebAssembly, introduces various ways to optimize Rust and WebAssembly with examples.

To get the most out of this book

The book assumes that you have a basic understanding of JavaScript. The code samples are mostly written in C++/Rust. Please install Rust (briefly explained in *Chapter 7, Integrating Rust with WebAssembly*) and Node.js before starting.

Software/hardware covered in the book	Operating system requirements
Rust	Windows, macOS, or Linux
Node.js >=14	

If you are using the digital version of this book, we advise you to type the code yourself or access the code from the book's GitHub repository (a link is available in the next section). Doing so will help you avoid any potential errors related to the copying and pasting of code.

Download the example code files

You can download the example code files for this book from GitHub at `https://github.com/PacktPublishing/Practical-WebAssembly`. If there's an update to the code, it will be updated in the GitHub repository.

We also have other code bundles from our rich catalog of books and videos available at `https://github.com/PacktPublishing/`. Check them out!

Conventions used

There are a number of text conventions used throughout this book.

`Code in text`: Indicates code words in text, database table names, folder names, filenames, file extensions, pathnames, dummy URLs, user input, and Twitter handles. Here is an example: "Mount the downloaded `WebStorm-10*.dmg` disk image file as another disk in your system."

A block of code is set as follows:

```
html, body, #map {
  height: 100%;
  margin: 0;
  padding: 0
}
```

When we wish to draw your attention to a particular part of a code block, the relevant lines or items are set in bold:

```
[default]
exten => s,1,Dial(Zap/1|30)
exten => s,2,Voicemail(u100)
exten => s,102,Voicemail(b100)
exten => i,1,Voicemail(s0)
```

Any command-line input or output is written as follows:

```
$ mkdir css
$ cd css
```

Bold: Indicates a new term, an important word, or words that you see onscreen. For instance, words in menus or dialog boxes appear in **bold**. Here is an example: "Select **System info** from the **Administration** panel."

> **Tips or Important Notes**
> Appear like this.

Get in touch

Feedback from our readers is always welcome.

General feedback: If you have questions about any aspect of this book, email us at customercare@packtpub.com and mention the book title in the subject of your message.

Errata: Although we have taken every care to ensure the accuracy of our content, mistakes do happen. If you have found a mistake in this book, we would be grateful if you would report this to us. Please visit www.packtpub.com/support/errata and fill in the form.

Piracy: If you come across any illegal copies of our works in any form on the internet, we would be grateful if you would provide us with the location address or website name. Please contact us at copyright@packt.com with a link to the material.

If you are interested in becoming an author: If there is a topic that you have expertise in and you are interested in either writing or contributing to a book, please visit authors.packtpub.com.

Share Your Thoughts

Once you've read *Practical WebAssembly*, we'd love to hear your thoughts! Scan the QR code below to go straight to the Amazon review page for this book and share your feedback.

https://packt.link/r/1838828001

Your review is important to us and the tech community and will help us make sure we're delivering excellent quality content.

Section 1: Introduction to WebAssembly

This section gives a brief introduction to LLVM and Emscripten. You will learn what they are and why you need to understand them before learning about WebAssembly. This section ends with an introduction to the WebAssembly module and WebAssembly text format.

You will understand what WebAssembly is and how it works after this section.

This section comprises the following chapters:

- *Chapter 1, Understanding LLVM*
- *Chapter 2, Understanding Emscripten*
- *Chapter 3, Exploring WebAssembly Modules*

1
Understanding LLVM

JavaScript is one of the most popular programming languages. However, JavaScript has two main disadvantages:

- **Unpredictable performance**

    JavaScript executes inside the environment and runtime provided by JavaScript engines. There are various JavaScript engines (V8, WebKit, and Gecko). All of them were built differently and run the same JavaScript code in a different way. Added to that, JavaScript is dynamically typed. This means JavaScript engines should guess the type while executing the JavaScript code. These factors lead to unpredictable performance in JavaScript execution. The optimizations for one type of JavaScript engine may cause undesirable side effects on other types of JavaScript engines. This leads to unpredictable performance.

- **Bundle size**

    The JavaScript engine waits until it downloads the entire JavaScript file before parsing and executing. The larger the JavaScript file, the longer the wait will be. This will degrade your application's performance. Bundlers such as webpack help to minimize the bundle size. But when your application grows, the bundle size grows exponentially.

Is there a tool that provides native performance and comes in a much smaller size? Yes, WebAssembly.

WebAssembly is the future of web and node development. WebAssembly is statically typed and precompiled, and thus it provides better performance than JavaScript. Precompilation of the binary provides an option to generate tiny binary bundles. WebAssembly allows languages such as Rust, C, and C++ to be compiled into binaries that run inside the JavaScript engine along with JavaScript. All WebAssembly compilers use LLVM underneath to convert the native code into WebAssembly binary code. Thus, it is important to understand what LLVM is and how it works.

In this chapter, we will learn what the various components of a compiler are and how they work. Then, we will explore what LLVM is and how it helps the compiled languages. Finally, we will see how the LLVM compiler compiles native code. We will cover the following topics in this chapter:

- Understanding compilers

- Exploring LLVM

- LLVM in action

Technical requirements

We will make use of **Clang**, which is a compiler that compiles C/C++ code into native code.

For Linux and Mac users, Clang should be available out of the box.

For Windows users, Clang can be installed from the following link: `https://llvm.org/docs/GettingStarted.html?highlight=installing%20clang%20windows#getting-the-source-code-and-building-llvm` to install Clang.

You can find the code files present in this chapter on GitHub at `https://github.com/PacktPublishing/Practical-WebAssembly`

Understanding compilers

Programming languages are broadly classified into compiled and interpreted languages.

In the compiled world, the code is first compiled into target machine code. This process of converting the code into binary is called *compilation*. The software program that converts the code into target machine code is called a *compiler*. During the compilation, the compiler runs a series of checks, passes, and validation on the code written and generates an efficient and optimized binary. A few examples of compiled languages are C, C++, and Rust.

In the interpreted world, the code is read and executed in a single pass. Since the compilation happens at runtime, the generated machine code is not as optimized as its compiled counterpart. Interpreted languages are significantly slower than compiled ones, but they provide dynamic typing and a smaller program size.

In this book, we will focus only on compiled languages.

Compiled languages

A compiler is a translator that translates source code into machine code (or in a more abstract way, converts the code from one programming language to another). A compiler is complicated because it should understand the language in which the source code is written (its syntax, semantics, and context); it should also understand the target machine code (its syntax, semantics, and context) and should create a representation that maps the source code into the target machine code.

A compiler has the following components:

- **Frontend** – The frontend is responsible for handling the source language.
- **Optimizer** – The optimizer is responsible for optimizing the code.
- **Backend** – The backend is responsible for handling the target language.

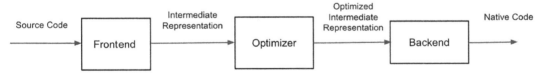

Figure 1.1 – Components of a compiler

Frontend

The frontend focuses on handling the source language. The frontend parses the code upon receiving it. The code is then checked for any grammar or syntax issues. After that, the code is converted (mapped) into an **intermediate representation** (**IR**). Consider IR as a format that represents the code that the compiler processes. The IR is the compiler's version of your code.

Optimizer

The second component in the compiler is the optimizer. This is optional, but as the name indicates, the optimizer analyzes the IR and transforms it into a much more efficient one. Few compilers have multiple IRs. The compiler efficiently optimizes the code on every pass over the IR. The optimizer is an IR-to-IR transformer. The optimizer analyzes, runs passes, and rewrites the IR. The optimizations here include removing redundant computations, eliminating dead code (code that cannot be reached), and various other optimizing options, which will be explored in future chapters. It is important to note that the optimizers need not be language-specific. Since they act on the IR, they can be built as a generic component and reused with multiple languages.

Backend

The backend focuses on producing the target language. The backend receives the generated (optimized) IR and converts it into another language (such as machine code). It is also possible to chain multiple backends that convert the code into some other languages. The backend is responsible for generating the target machine code from the IR. This machine code is the actual code that runs on the bare metal. In order to produce efficient machine code, the backend should understand the architecture in which the code is executed.

Machine code is a set of instructions that instructs the machine to store some values in registers and do some computation on them. For example, the generated machine code is responsible for efficiently storing a 64-bit number in 32-bit architecture in a free register (and things like that). The backend should understand the target environment to efficiently create a set of instructions and properly select and schedule the instructions to increase the performance of the application execution.

Compiler efficiency

The faster the execution, the better the performance.

The efficiency of the compiler depends on how it selects the instruction, allocates the register, and schedules the instruction execution in the given architecture. An instruction set is a set of operations supported by a processor, and this overall design is called an **Instruction Set Architecture (ISA)**. The ISA is an abstract model of a computer and is often referred to as computer architecture. Various processors convert the ISA in different implementations. The different implementations may vary in performance. The ISA is an interface between the hardware and the software.

If you are implementing a new programming language and you want this language to be running on different architectures (or, more abstractly, different processors), then you should build the backend for each of these architectures/targets. But building these backends for every architecture is difficult and will take time, cost, and effort to embark on a language creation journey.

What if we create a common IR and build a compiler that converts this IR into machine code that runs efficiently on various architecture? Let's call this compiler a low-level virtual machine. Now, the role of your frontend in the compiler chain is just to convert the source code into an IR that is compatible with a low-level virtual machine (such as LLVM). Now, the general purpose of a low-level virtual machine is to be a common reusable component that maps the IR into native code for various targets. But the low-level virtual machine will only understand the common IR. This IR is called the **LLVM IR** and the compiler is called **LLVM**.

Exploring LLVM

LLVM is a part of the LLVM Project. The LLVM Project hosts compilers and toolchain technologies. The *LLVM core* is a part of the LLVM Project. The LLVM core is responsible for providing source- and target-independent optimization and for generating code for many CPU architectures. This enables language developers to just create a frontend that generates an LLVM-compatible IR or LLVM IR from the source language.

Did You Know?

LLVM is not an acronym. When the project was started as a research project, it meant Low-Level Virtual Machine. But later, it was decided to use the name as it is rather than as an acronym.

The main advantages of LLVM are as follows:

- LLVM uses a simple low-level language that looks similar to C.

- LLVM is strongly typed.

- LLVM has strictly defined semantics.

- LLVM has accurate and precise garbage collection.

- LLVM provides various optimizations that you can choose based on the requirement. It has *aggressive, scalar, inter-procedural, simple-loop*, and *profile-driven* optimizations.

- LLVM provides various compilation models. They are *link time*, *install time*, *runtime*, and *offline*.

- LLVM generates machine code for various target architectures.

- LLVM provides DWARF debugging information.

> **Note**
>
> DWARF is a debugging file format used by many compilers and debuggers to support source-level debugging. DWARF is architecture-independent and applicable to any processor or operating system. It uses a data structure called a **Debugging Information Entry (DIE)** to represent each variable, type, procedure, and so on.
>
> If you want to explore more about DWARF, refer to `http://dwarfstd.org/doc/Debugging%20using%20DWARF-2012.pdf`.

> **Important Note**
>
> LLVM is not a single monolithic project. It is a collection of subprojects and other projects. These projects are used by various languages, such as Ruby, Python, Haskell, Rust, and D, for compilation.

Now that we have an understanding of compilers and LLVM, we will see how it is used.

LLVM in action

In this section, let's use LLVM's Clang compiler to compile native code into LLVM IR. This will give a better idea of how LLVM works and will be useful for understanding how the compilers use LLVM in future chapters.

We first create a C file called `sum.c` and enter the following contents:

```
$ touch sum.c
// sum.c
unsigned sum(unsigned a, unsigned b) {
    return a + b;
}
```

The sum.c file contains a simple sum function that takes in two unsigned integers and returns the sum of them. LLVM provides the Clang LLVM compiler to compile the C source code. In order to generate the LLVM IR, run the following command:

```
$ clang -S -O3 -emit-llvm sum.c
```

We provided the Clang compiler with the -S, -O3, and -emit-llvm options:

- The -S option specifies for the compiler to only run the preprocess and compilation steps.
- The -O3 option specifies for the compiler to generate a well-optimized binary.
- The -emit-llvm option specifies for the compiler to emit the LLVM IR while generating the machine code.

The preceding code will print out the following LLVM IR:

```
define i32 @sum(i32, i32) local_unnamed_addr #0 {
  %3 = add i32 %1, %0
  ret i32 %3
}
```

The syntax of the LLVM IR is structurally much closer to C. The define keyword defines the beginning of a function. Next to that is the return type of the function, i32. Next, we have the name of the function, @sum.

> **Important Note**
> Note the @ symbol there? LLVM uses @ to identify the global variables and function. It uses % to identify the local variables.

After the function name, we state the types of the input argument (i32 in this case). The local_unnamed_addr attribute indicates that the address is known not to be significant within the module. The variables in the LLVM IR are *immutable*. That is, once you define them, you cannot change them. So inside the `block`, we create a new local value, %3, and assign it the value of add. add is an opcode that takes in the `type` of the arguments followed by the two arguments, %0 and %1. %0 and %1 denote the first and second local variables. Finally, we return %3 with the ret keyword followed by the `type`.

This IR is transformable; that is, the IR can be transformed from the textual representation into memory and then into actual bit code that run on the bare metal. Also, from bit code, you can transform them back to the textual representation.

Imagine that you are writing a new language. The success of the language depends on how versatile the language is at performing on various architectures. Generating optimized byte codes for various architectures (such as x86, ARM, and others) takes a long time and it is not easy. LLVM provides an easy way to achieve it. Instead of targeting the different architecture, create a compiler frontend that converts the source code into an LLVM compatible IR. Then, LLVM will convert the IR into efficient and optimized byte code that runs on any architecture.

> **Note**
>
> LLVM is an umbrella project. It has so many components that you could write a set of books on them. Covering the whole of LLVM and how to install and run them is beyond the scope of this book. If you are interested in learning more about various components of LLVM, how they work, and how to use them, then check out the website: `https://llvm.org`.

Summary

In this chapter, we have seen how compiled languages work and how LLVM helps to compile them. We have compiled a sample program with LLVM to understand how it works. In the next chapter, we'll explore Emscripten, a tool that converts C/C++ into a WebAssembly module. Emscripten uses the LLVM backend to do the compilation.

2
Understanding Emscripten

In this chapter, we will learn about **Emscripten**, which is a toolchain to convert C/C++ code into a WebAssembly module.

Emscripten consists of two components:

- Emscripten compiler frontend
- **Emscripten SDK (emsdk)**

The **Clang** compiler frontend compiles C/C++ code into **LLVM intermediate representation** (**LLVM IR**) and then uses the LLVM backend to convert the LLVM IR into native code. The Clang compiler is fast, uses little memory, and is compatible with **GNU Compiler Collection** (**GCC**). Emscripten is similar to Clang; the former produces a *wasm* binary while the latter produces a *native* binary. The **Emscripten compiler frontend** (**emcc**) is the compiler frontend that converts C/C++ into LLVM IR (both binary and human-readable form) and into the WebAssembly binary or *asm.js*, such as JavaScript.

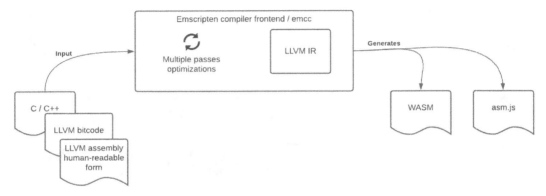

Figure 2.1 – Emscripten compiler frontend

emsdk helps manage and maintain the Emscripten toolchain components and set up the runtime/terminal environment to run emcc.

In this chapter, we will learn how to install Emscripten. Then, we will use Emscripten to generate asm.js, a WebAssembly module that runs on Node.js and the browser. After that, we will explore the emsdk tool. Finally, we will explore various optimizations provided by Emscripten. We will cover the following topics in this chapter:

- Installing Emscripten using emsdk

- Generating asm.js using Emscripten

- Running Hello World with Emscripten in Node.js

- Running Hello World with Emscripten in the browser

- Exploring other options in emsdk

- Understanding various levels of optimizations

> **Did You Know?**
> asm.js is a subset of JavaScript that is optimized to run at near-native performance in the browser. The asm.js spec was not accepted by all browser vendors. asm.js has evolved into WebAssembly.

Technical requirements

We will be showing how to set up Emscripten in this chapter, for which you will require the following installed on your system:

- Python >= 3.7

- Node.js > 12.18

> **Note**
> emsdk comes prebundled with a compatible Node.js version.

You can find the code files present in this chapter on GitHub at https://github.com/PacktPublishing/Practical-WebAssembly

Installing Emscripten using emsdk

emsdk provides an easy way to install, manage, and switch versions of the Emscripten toolchain. emsdk takes care of setting up the environment, tools, and SDK required for compiling C/C++ to LLVM IR and then to JavaScript in the form of asm.js or the WebAssembly binary.

Let's install Emscripten and start hacking:

1. Clone the emsdk repo and go into the emsdk folder:

```
$ git clone https://github.com/emscripten-core/emsdk
$ cd emsdk
```

2. To install emsdk on the machine, run the following command:

For *nix users, use the following:

```
$ ./emsdk install latest
```

For Windows users, use the following:

```
$ emsdk install latest
```

> **Note**
> The preceding command might take a while to run; it will build and set up the entire toolchain.

Next, we will activate the latest emsdk. The activation updates the local shell with the necessary environment references and makes the latest SDK active for the user in the current shell. It writes the path and other necessary information for the Emscripten toolchain to work under the user's home directory in a file called .emscripten:

3. To activate the installed emsdk, run the following command:

    For *nix users, use the following:

    ```
    $ ./emsdk activate latest
    ```

    For Windows users, use the following:

    ```
    $ emsdk activate latest
    ```

4. Now it is time to make sure the configurations and paths are activated by running the following:

    For *nix users, use the following:

    ```
    $ source ./emsdk_env.sh
    ```

    For Windows users, use the following:

    ```
    $ emsdk_env.bat
    ```

Congrats, the Emscripten toolchain is installed! Updating the toolchain with emsdk is as easy as installing it.

To update, run the following command:

For *nix users, use the following:

```
$ ./emsdk update
```

For Windows users, use the following:

```
$ emsdk update
```

emsdk sets up the following paths inside the Emscripten configuration file. The Emscripten configuration file (.emscripten) is in the home folder. It consists of the following:

- LLVM_ROOT – specifies the path of the LLVM Clang compiler
- NODE_JS – specifies the path of Node.js
- BINARYEN_ROOT – specifies the optimizer for the Emscripten compiler
- EMSCRIPTEN_ROOT – specifies the path of the Emscripten compiler

We can check whether the installation of emcc was successful by using the following command:

```
$ emcc --version
```

Now that we have finished installing the Emscripten compiler, let's go ahead and use it.

Generating asm.js using Emscripten

We will use Emscripten to port C/C++ programs into asm.js or the WebAssembly binary and then run them inside the JavaScript engine.

> **Note**
>
> Programming languages such as Lua and Python have a C/C++ runtime. With Emscripten, we can port the runtime as a WebAssembly module and execute them inside the JavaScript engine. This makes it easy to run Lua/Python code on the JavaScript engine. Thus, Emscripten and WebAssembly allow the running of native code in the JavaScript engine.

First, let's create a sum.cpp file:

```
// sum.cpp
extern "C" {
  unsigned sum(unsigned a, unsigned b) {
      return a + b;
  }
}
```

Consider extern "C" as something like an *export* mechanism. All the functions inside are available as an exported function without any changes to their name. Then, we define the normal sum function that takes in two numbers and returns a number.

In order to generate the asm.js like JavaScript code from sum.cpp, use the following command:

```
$ emcc -O1 ./sum.cpp -o sum.html -sWASM=0 -sEXPORTED_
  FUNCTIONS='["_sum"]'
```

> **Note**
>
> If you are running emcc for the first time, it might take a few seconds to complete. Subsequent runs will be faster.

We pass in the `-O1` option to the emcc compiler, instructing the compiler to produce less-optimized code (we will see more options to optimize later in this chapter). Next, we pass the file to be converted, that is, `sum.cpp`. Then, with the `-o` flag, we provide the desired name for the output, which is `sum.html`.

Finally, we send in more information to the emcc compiler using the `-s` flag. The `-s` flag takes in a key and value as their arguments. The emcc compiler generates the WebAssembly module by default. `WASM=0` instructs the compiler to generate asm.js like JavaScript instead of WebAssembly.

Then, we specify the exported functions using the `EXPORTED_FUNCTIONS` option. The `EXPORTED_FUNCTIONS` option takes an array of arguments. In order to export the `sum` function, we specify `_sum`.

This will generate the following code:

```
/// $1 is mapped to _sum

function $1($0_1, $1_1) {

    $0_1 = $0_1|0;

    $1_1 = $1_1|0;

    return $0_1 + $1_1 | 0;

}
```

> **Note**
> | 0 specifies the type as a number.

Now open `sum.html` in a browser and open the developer console. In order to call the exported function, we will run the following expression in the console:

```
ccall("sum", "number", "number, number", [10, 20])
// outputs 30
```

`ccall` is the way to call the exported function from C/C++ code via JavaScript. The function takes in the name of the function, the type of the return value, types of arguments, and then the input arguments as an array. This will invoke the `sum` function to produce the result. We will see more about `ccall` and `cwrap` in later chapters. But for now, consider `ccall` a way to call the C function.

Find out more about the Emscripten source at `https://github.com/emscripten-core/emscripten`.

So far, we have seen how to generate asm.js files using emscripten. Let us use emscripten to create a WebAssembly Module to run on Node.js.

Running Hello World with Emscripten in Node.js

In this section, we will see how to convert C/C++ code into the WebAssembly binary via Emscripten and run it along with Node.js.

> **Note**
>
> If the terminal errors out with *emcc command not found*, your terminal environment might have been reset. To set up the environment, run the following command from inside the `emsdk` folder:
>
> `source ./emsdk_env.sh`

Let's follow the tradition of Brian Kernighan, by writing "Hello, world" with a slight twist. Let's do a "Hello, Web":

1. First, we create a `hello_web.c` file:

   ```
   $ touch hello_web.c
   ```

2. Launch your favorite editor and add the following code:

   ```
   #include <stdio.h>

   int main() {
       printf("Hello, Web!\n");
       return 0;
   }
   ```

It is a simple C program with a `main` function. The `main` function is the entry point during the runtime. When this code is compiled and executed using Clang (`clang sum.c && ./a.out`), "Hello, Web!" is printed. Now, instead of Clang (or any other compiler), let's compile the code with emcc.

3. We enter the following command to compile the code with emcc:

```
$ emcc hello_web.c
```

Once completed, the following files are generated:

- a.out.js

- a.out.wasm

The generated JavaScript file is huge. It has more than 2,000 lines and is 109 KB in size. We will learn how to optimize the file size later in this chapter.

4. Let's run the generated JavaScript file using Node and that will print out "Hello, Web!":

```
$ node a.out.js
Hello, Web!
```

Congratulations! You just ran your first WebAssembly binary!

> **Note**
> Binary size matters in the world of browsers. It won't matter if your algorithm runs in nanoseconds if you have a huge chunk of code. The browser waits till it receives all the necessary information before it starts to parse and compile. So, it is mandatory to check the file size. *Closure Compiler* helps to minimize the byte code size further. Closure Compiler not only reduces the code size but also tries to make the code more efficient.

The generated JavaScript file contains its own runtime and configuration needed for the JavaScript engine to execute the WebAssembly module inside the JavaScript engine. The generated JavaScript file creates a JavaScript module and initializes code for both browsers and Node.js:

- In Node.js, the generated JavaScript file creates a module by reading the file from the local filesystem. It gets the arguments passed to the node command and sets them up in the module created.

- In browsers, the generated JavaScript file creates a module by framing a request and fetches it as bytes from a URL. The browser fetches the WebAssembly binary from the hosted server or location and then instantiates the module.

The generated JavaScript file also creates a stack, memory, import, and export sections. We will deep dive into those sections later in this book.

This generated JavaScript file is called a *binding file*. The main function of the binding file is to create or set an environment that enables executing a WebAssembly module inside the JavaScript engine. The binding file acts as a translator between JavaScript and WebAssembly. All the values are passed in and out via this binding file.

When the JavaScript file is executed via node, it does the following.

The JavaScript engine first loads the module and then sets up the constants and various functions that are required for WebAssembly to execute. Then, the module checks where the code is being executed, whether the module is inside the browser or in the `Node` environment. Based on that, it fetches the file. Since we are running the WebAssembly Module via the node here, it fetches the file from the local filesystem. Then, the module is checked for any arguments provided for the call. If not, the JavaScript engine will check whether there are any unhandled/uncaught exceptions. The JavaScript engine then maps the `print out`/`print err` function to the console. The JavaScript engine checks whether the module loaded has all the required access and global variables and imports are available for execution.

The module goes on to initialize the stack and other required constants as well as the decoder and encoder for decoding and encoding the buffer, respectively. The encoder is responsible for translating the JavaScript values into WebAssembly-understandable values. The decoder is responsible for translating the WebAssembly values into JavaScript-understandable values.

The Node.js runtime then checks the availability of the file and then initializes the file. The module is checked for all the WebAssembly-related function availability. Once everything is initialized and the module contains all the functions required, we will call the `run` function.

The `run` function instantiates the WebAssembly binary. In this case, since we have defined the `main` function in C, the binding file calls the `main` function straight away when instantiated.

The binding file contains the `ccall` function. The `ccall` function is an interface to the underlying function defined in C:

```
function ccall(ident, returnType, argTypes, args, opts) {
    // the code is elided
}
```

The `ccall` function accepts the following arguments:

- `ident` – The function to call; it is the function identifier defined in C.
- `returnType` – The return type of the function.
- `argTypes` – The argument types.
- `args` – The arguments that are passed along with the function call.
- `opts` – Any other options that are required.

The JavaScript module exports the `cwrap` function in addition to `ccall`. `cwrap` is a wrapper function around the `ccall` function. While `ccall` is a function invocation, `cwrap` provides a function that invokes `ccall`:

```
function cwrap(ident, returnType, argTypes, opts) {
    return function() {
        return ccall(ident, returnType, argTypes,
            arguments, opts);
    }
}
```

The WebAssembly file generated consists of binary opcode to instruct the runtime to print "Hello, Web!". The WebAssembly file starts with `00 61 73 6d 01 00 00 00`.

Find out more about WebAssembly specifications at `https://webassembly.github.io/spec/`.

So far, we have seen how to generate a WebAssembly module to run on Node.js. Let us use emscripten to create a WebAssembly Module to run in the browser.

Running Hello World with Emscripten in the browser

In this section, we will see how to convert C/C++ code into the WebAssembly binary via Emscripten and run it in the browser.

> **Note**
>
> If the terminal says that the `emcc` command is not found, it is highly likely that you have missed setting up the environment variables. To set up the environment variables, run the following command from inside the `emsdk` folder: `source ./emsdk_env.sh`

Let's use the same code example used in the *Generating asm.js using Emscripten* section . Now, instead of just running emcc, let's pass the `-o` option and instruct emcc to generate the `.html` file:

```
$ emcc hello_web.c -o helloweb.html
```

Once completed, the following files are generated:

- `helloweb.js`
- `helloweb.wasm`
- `helloweb.html`

Similar to the Node example, the generated JavaScript file is huge. We will learn how to optimize the file size later in this chapter.

> **Note**
>
> The `-o` option ensures all the files generated have the name `helloweb`.

In order to run the generated HTML file in the browser, we will need a web server. The web server serves the HTML file over the HTTP protocol. Explaining web servers and how they work is beyond the scope of this book; refer to `https://en.wikipedia.org/wiki/Web_server` for more details.

Python provides an easy way to run the web server. In order to run the web server using Python, run the following command:

```
$ python -m http.server <port number>
```

Open `http://localhost:<port number>/helloweb.html` to see WebAssembly in action in the browser.

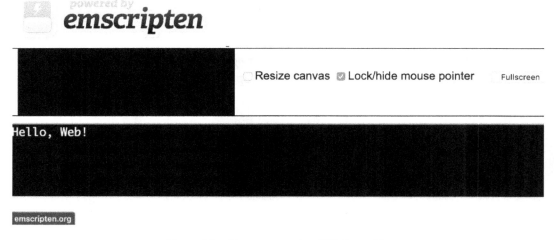

Figure 2.2 – Browser running WebAssembly

When the JavaScript file is executed via the browser, it prints out **Hello, Web!** as output. The only difference is instead of loading the WASM file from the filesystem, it loads it via XHR. Once everything is initialized and the module consists of all the functions required, we will call the `run` function. The `run` function instantiates the WebAssembly binary. In this case, since we have defined the `main` function in C, the binding file calls the `main` function straight away when instantiated.

Emscripten also provides `emrun` to run HTML files. Check out `https://emscripten.org/docs/compiling/Running-html-files-with-emrun.html` for more information.

Find out more about deploying Emscripten-compiled pages at `https://emscripten.org/docs/compiling/Deploying-Pages.html`.

We have used Emscripten to generate the WebAssembly module. Let's go ahead and explore what else the emsdk can do.

Exploring other options in emsdk

emsdk is a single-stop shop for installing, maintaining, and managing all the tools and toolchains required for using Emscripten. emsdk makes it easier to bootstrap the environment, upgrade to the latest versions, switch to various versions, change or configure various tools, and so on.

The emsdk command is available inside the emsdk folder. Go to the emsdk folder and run the emsdk command.

> **Note**
>
> For all the commands in this chapter, for *nix systems, use ./emsdk, and for Windows, use emsdk.

To find the various options available in the emsdk command, run the following command:

```
$ ./emsdk --help
emsdk: Available commands:
 emsdk list [--old] [--uses] - To list down the tools
 emsdk update - To update the emsdk to the latest version.
 emsdk update-tags - To fetch the latest tags from the GitHub
   repository.
 emsdk install - To install the tools and SDK.
 emsdk uninstall - To uninstall the tools and SDK installed
   previously.
 emsdk activate - To activate the currently installed version.
```

An emsdk command takes the following format:

```
emsdk <option> <Tool / SDK > --<flags>
```

The emsdk command consists of the following:

- <option>

  This can be one of the following: list, update, update-tags, install, uninstall, or activate.

- <Tool/SDK>

  This refers to libraries and it includes Emscripten and LLVM. SDK refers to emsdk itself.

- --<flags>

  This refers to various configuration options.

Let us explore each of the options and flags that emsdk command supports.

Listing the tools and SDK

Here, we show how to list the tools and SDK that are available with emsdk. Run the following command:

```
$ ./emsdk list

The *recommended* precompiled SDK download is 2.0.6
   (4ba921c8c8fe2e8cae071ca9889d5c27f5debd87).

To install/activate it, use one of:
        latest                    [default (llvm) backend]
        latest-fastcomp           [legacy (fastcomp) backend]

Those are equivalent to installing/activating the following:
        2.0.6               INSTALLED
        2.0.6-fastcomp

All recent (non-legacy) installable versions are:
        2.0.6     INSTALLED
        ...

The additional following precompiled SDKs are also available
   for download:
        sdk-fastcomp-1.38.31-64bit

The following SDKs can be compiled from source:
        sdk-upstream-master-64bit
        ...
The following precompiled tool packages are available for
   download:
        ...
    *       node-12.18.1-64bit              INSTALLED
    *       python-3.7.4-2-64bit            INSTALLED
            emscripten-1.38.30
        ...
The following tools can be compiled from source:
```

```
        llvm
        clang
        emscripten
        binaryen


Items marked with * are activated for the current user.


To access the historical archived versions, type 'emsdk list
   --old'


Run "git pull" followed by "./emsdk update-tags" to pull
   in the latest list.
```

emsdk list lists all tool packages and SDKs that are available. This list of tools and SDKs includes the last few versions of LLVM, Clang, Emscripten, and Binaryen. They even have Node versions 8 and 12 and Python 3.7. emsdk maintains and manages emsdk. This means that we need to know the information about the current version that we are using and how to update it. The emsdk list command also provides more detail on the SDK components along with a list of those compiled from the sources.

Managing the tools and SDK

emsdk provides an option to install, update, and uninstall the tools and SDK.

In order to install the tools, SDK, or emsdk itself, use the following:

```
$ ./emsdk install <tool / SDK to install>
```

To install the latest version of the SDK, you can run the following:

```
./emsdk install latest
```

> **Note**
> latest refers to the latest version of emsdk.

To install multiple tools with the emsdk install command, use the following:

```
./emsdk install <tool1> <tool2> <tool3>
```

You can also specify multiple options for the `install` command. You can pass in options to the `install` command like this:

```
./emsdk install [options] <tools / SDK>
```

The various `options` available are as follows:

- Number of cores to build
- Type of build
- Activation of tools and SDK
- Uninstallation

Number of cores to build

The initial setup will take a long time to build and install the required tools and SDK. Based on your requirements, you can control the number of cores that you need for building and installing the required tools and SDK:

```
./emsdk install -j<number of cores to use for building> <tools
  / SDK>
```

Type of build

You instruct `emsdk` on what type of build is to be used to make LLVM perform:

```
./emsdk install --build=<type> <tools / SDK>
```

`type` accepts the following options:

- Debug
    - This type is used for debugging.
    - It generates Symbol files.
    - The end build will not produce optimized, fast code.
- Release
    - This type will generate optimized, fast code.

- MinSizeRel

  - This type is the same as Release.

  - This type will minimize the size and maximize the speed.

  - This uses optimization options such as -O1 (minimize size) and -O2 (maximize speed).

- RelWithDebInfo

  - This type is the same as Release.

  - This type will also generate Symbol files. That will help in debugging.

Activation of the tools and SDK

After the tools and SDK are installed, we can activate different versions to use them. The activate command generates the necessary configuration files mapping the path, with the built executables.

To activate the tools and SDK, run the following:

```
./emsdk activate <tools / SDK to activate>
```

The activate command accepts a few options; they are as follows:

- --embedded – This option makes sure all the built files, configuration, cache, and temporary files are located inside the directory in which the emsdk command is located.

  If not specified, this command will move the configuration file to the user's home directory.

- --build=<type> – Similar to the type of build that LLVM supports. For example, Debug, Release, MinSizeRel, RelWithDebInfo.

Uninstallation of the tools and SDK

To uninstall the tools and SDK, we can run the following:

```
./emsdk uninstall < tools / SDK to uninstall>
```

Find out more about the tools at https://emscripten.org/docs/tools_reference/index.html.

We have explored how emsdk helps us to manage tools and SDKs; let's go ahead and explore various optimizations provided by Emscripten.

Understanding various levels of optimizations

C/C++ programs are compiled and converted into native code via Clang or the GCC compiler. Clang or the GCC compiler converts the C/C++ program based on the target. Target here refers to the end machine where the code is executed. emcc has the Clang compiler built in. The emcc compiler is responsible for converting the C or C++ source code into LLVM byte code.

In this section, we will see how to improve the optimization and code size of the generated WebAssembly binary code.

To improve the efficiency and generated code size, the Emscripten compiler has the following options:

- Optimizations
- Closure Compiler

Lets talk about optimizations first.

Optimizations

The goal of the compiler is to reduce the cost of compilation, that is, the compile time. With the -O optimization flag, the compiler tries to improve the code size and/or the performance at the expense of the compile time. In terms of compiler optimizations, code size and performance are mutually exclusive. The faster the compile time, the lower the optimization. To specify the optimization, we use the -O<0/1/2/3/s/z> flag. Each of the options includes various assertions, code size optimizations, and code performance optimizations, along with others.

The following are the various optimizations available:

- -O0 – This is the default option and a perfect starter to experiment. This option means "no optimizations." This optimization level compiles the fastest and generates the most debuggable code. This is a basic optimization level. This option tries to inline functions.

- -O1 – This option adds simple optimizations and tries to generate a minimum code size. This option removes runtime assertions in the generated code and builds slower than the -O0 option. This option also tries to simplify the loops.

- -O2 – This option adds further optimizations than -O1. It is slower than -O1 but generates code that is more optimized than the -O1 option. This option optimizes the code based on JavaScript optimization and removes code that is not part of JavaScript modules. This option removes the inline functions and the vectorize-loop option is set. This option adds a moderate level of optimization. This option also adds dead code elimination.

  Vectorization will instruct the processor to do the operation in chunks rather than doing it one by one.

- -O3 – This option adds more options, takes more time to compile, and generates more optimized code than the -O2 option.

  This option produces optimal production-ready code. This option is like -O2, except that it enables optimizations that take longer to perform or that may generate larger code (in an attempt to make the program run faster).

- -Os – This option is similar to -O2. It adds extra optimizations and reduces the code size. Reducing the code size in turn decreases the performance. This option generates smaller code than -O2.

- -Oz – This option is similar to -Os but reduces the code size even further. This option takes more compile time to generate the binary code.

We will now explore the various optimization options provided by Emscripten:

1. First, we create a C file called optimization_check:

   ```
   $ touch optimization_check.c
   ```

2. Then, open your favorite editor and add the following code. The following is a simple C file with a main function and a couple of other functions:

   ```c
   #include <stdio.h>
     int addSame(int a) {
        return a + a;
   }

   int add(int a, int b) {
        return a + b;
   }

   int main() {
        printf("Hello, Web!\n");
   ```

```
    int a;
    int sum = 0;
    /* for loop execution */
    for( a = 0; a < 20; a = a + 1 ){
        sum = sum + a;
    }
    addSame(sum);
    add(1, 2);
    return 0;
}
```

3. We then compile this into WebAssembly code using emcc:

```
$ time emcc optimization_check.c
emcc optimization_check.c   0.32s user 0.14s system
   90% cpu 0.514 total
```

4. Then, check the sizes of the file generated:

```
$ l
324B optimization_check.c
13K a.out.wasm
109K a.out.js
```

We can see that the WebAssembly file generated is about 13 KB and it took a total of 0.514 seconds to compile. That is a fast compilation but the code size is huge.

In the world of compilers, the faster the compilation, the bigger the code size and the slower the execution speed will be.

5. Now, let's optimize it further using the -O1 option:

```
$ time emcc -O1 optimization_check.c
emcc -O1 optimization_check.c   0.31s user 0.13s system
   86% cpu 0.519 total
```

Check the sizes of the file generated:

```
$ l
324B optimization_check.c
3.4K a.out.wasm
59K a.out.js
```

The WebAssembly file generated is about 3.4 KB (3.8 times less than the -OO version) and it took almost the same time, around 0.519 seconds.

6. Now, let's optimize it further using the -O2 option:

```
$ time emcc -O2 optimization_check.c
emcc -O2 optimization_check.c   0.53s user 0.16s system
   111% cpu 0.620 total
```

Check the sizes of the file generated:

```
$ l
324B optimization_check.c
2K a.out.wasm
20K a.out.js
```

The WebAssembly file generated is about 2 KB (~6.5 times less than -OO) and it took around 0.62 seconds.

7. Now, let's optimize it further using the -O3 option:

```
$ time emcc -O3 --profiling optimization_check.c
emcc -O3 --profiling optimization_check.c   1.03s user
   0.21s system 110% cpu 1.117 total
```

Find out more about the --profiling flag at https://emscripten.org/docs/tools_reference/emcc.html#emcc-profiling.

Check the sizes of the file generated:

```
$ l
324B optimization_check.c
2.0K a.out.wasm
17K a.out.js
```

The WebAssembly file generated is the same size as -O2 but the generated JavaScript file is 3 KB less, and it took around 1.117 seconds to compile.

8. Now, let's optimize it further using the -Os option:

```
$ time emcc -Os optimization_check.c
emcc -Os optimization_check.c   1.03s user 0.22s system
   46% cpu 2.655 total
```

Check the sizes of the file generated:

```
$ l
324B optimization_check.c
1.7K a.out.wasm
14K a.out.js
```

The WebAssembly file generated is about 1.7 KB (~7.5 times less than -O0) and it took almost 2.655 seconds.

9. Now, let's optimize it further using the -Oz option:

```
$ time emcc -Oz optimization_check.c
emcc -Oz optimization_check.c  1.03s user 0.21s system
   110% cpu 1.123 total
```

Check the sizes of the file generated:

```
$ l
324B optimization_check.c
1.7K a.out.wasm
14K a.out.js
```

The WebAssembly file generated is about 1.7 KB (~7.5 times less than -O0) and it took around 1.123 seconds.

Next, we'll see an alternative means provided by the Emscripten compiler for improving the efficiency and reducing the generated code size: Closure Compiler

Closure Compiler

Closure Compiler is a tool for compiling JavaScript to better JavaScript. It parses, analyzes, removes dead code, rewrites, and minimizes JavaScript. Further optimizations on the generated binding JavaScript file and WebAssembly module are done using Closure Compiler. With Closure Compiler, we can do better optimizations on Emscripten code. To optimize the WebAssembly module and JavaScript further, we can use --closure <optimization type>.

The optimization type has the following options:

- --closure 0 – This option adds no Closure Compiler optimizations.

- --closure 1 – This option reduces the generated JavaScript code size. This option does not optimize the asm.js and WebAssembly binary. This adds an additional compilation step that increases the compilation time.

- `--closure 2` – This option optimizes JavaScript, asm.js, and not the WebAssembly binary and reduces the code size of the file drastically for asm.js.

We will use the `-closure 1` option to optimize the WebAssembly binary along with the `-O3/s` Emscripten optimization options:

```
$ time emcc -O3 --closure 1 optimization_check.c
emcc -O3 --closure 1 optimization_check.c  2.40s user 0.42s
  system 105% cpu 2.681 total
```

The file sizes generated are as follows:

```
$ l
324B optimization_check.c
1.8K a.out.wasm
6.5K a.out.js
```

Along with `emcc -O3`, we pass `-closure 1` to further optimize the generated files. Closure Compiler reduces the size of the JavaScript file by 50% when compared to the `emcc -O3` option and it took 2.681 seconds to compile:

```
time emcc -Os --closure 1 optimization_check.c
emcc -Os --closure 1 optimization_check.c  2.53s user 0.42s
  system 106% cpu 2.778 total
```

Let's list the files in the current folder to check the generated files and their size:

```
$ l
324B optimization_check.c
1.7K a.out.wasm
6.5K a.out.js
```

Along with `emcc -Os`, we pass `-closure 1` to further optimize the generated binary. Closure Compiler reduces the `.wasm` file a little bit more with the `emcc -Os` option, and it took 2.778 seconds to compile.

> **Note**
>
> When optimizing for size, try to use both `-O3` or `-Os` along with `--closure 1` to optimize both JavaScript and the WebAssembly module.

Check out more various options and flags at `https://emscripten.org/docs/tools_reference/emcc.html` `https://clang.llvm.org/docs/CommandGuide/clang.html`.

Find out more about various available optimization options at `https://docs.microsoft.com/en-us/cpp/build/reference/o-options-optimize-code?view=vs-2017`.

Learn more about Closure Compiler at `https://developers.google.com/closure/compiler`.

Find out more about optimizing large code bases with Emscripten at `https://emscripten.org/docs/optimizing/Optimizing-Code.html#very-large-codebases`.

Summary

In this chapter, we learned how to install and use Emscripten to compile C/C++ into a WebAssembly module. We also explored the emsdk tool and various levels of optimizations when generating the WebAssembly module. In the next chapter, we will explore the WebAssembly module.

3
Exploring WebAssembly Modules

WebAssembly is a low-level assembly-like code that is designed for efficient execution and compact representation. WebAssembly runs at a near-native speed in all JavaScript engines (including modern desktop and mobile browsers and Node.js). Compact representation of the binary enables the generated binary to be as small as possible in size.

> **Note**
> The main goal of WebAssembly is to enable high-performance applications.

Each WebAssembly file is an efficient, optimal, and self-sufficient module called a **WebAssembly module (WASM)**. WASM is safe, that is, the binary runs in a memory-safe and sandboxed environment. WASM does not have permission to access anything outside of that sandbox. WASM is language-, hardware-, and platform-independent.

WebAssembly is a virtual **instruction set architecture (ISA)**. The WebAssembly specifications define the following:

- Instruction set
- Binary encoding
- Validation
- Execution semantics

The WebAssembly specification also defines a textual representation of the WebAssembly binary.

In this chapter, we will explore WASM and how a JavaScript engine executes WASM. We then explore the WebAssembly text format and why it is useful. Understanding WASM execution and the WebAssembly text format will enable us to easily understand the module and debug it in the JavaScript engine. We will cover the following main topics in this chapter:

- Understanding how WebAssembly works
- Exploring the WebAssembly text format

Technical requirements

You can find the code files present in this chapter on GitHub at `https://github.com/PacktPublishing/Practical-WebAssembly`.

Understanding how WebAssembly works

Let's first explore how JavaScript and WebAssembly are executed inside the JavaScript engine.

Understanding JavaScript execution inside the JavaScript engine

The JavaScript engine first fetches the complete JavaScript file (note that the engine has to wait until the entire file is downloaded/loaded).

Note

The bigger the JavaScript file, the longer it takes to load. It doesn't matter how fast your JavaScript engine is or how efficient your code is. If your JavaScript file is huge (that is, greater than 170 KB), then your application is going to be slow at loading time.

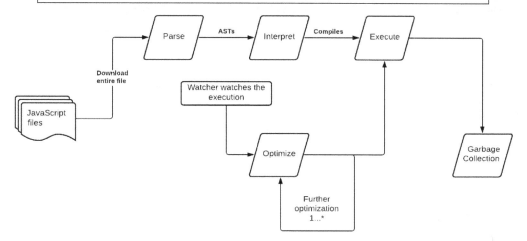

Figure 3.1 – JavaScript execution inside the JavaScript engine

Once loaded, the JavaScript is parsed into **abstract syntax trees** (**ASTs**). This phase is called **parse**. Since JavaScript is both an interpreted and compiled language, the JavaScript engine kickstarts the execution after parsing. The interpreter executes the code faster but it compiles the code every time. This phase is called **interpret**.

The JavaScript engine has **watchers** (called **profilers** in some browsers). Watchers keep track of code execution. If a particular block of code is executed frequently, then the watcher marks it as hot code. The engine compiles the block of code using the **just-in-time** (**JIT**) compiler. The engine spends some time doing the compilation, say in the order of nanoseconds. The time spent here is worth it, because the next time the function is called, the execution happens much faster, because the compiled version is always faster than the interpreted one. This phase is called **optimize**.

JavaScript engines add one (or two) more layers of optimization. The watchers continue watching the code execution. The watchers then name the code that is called more often *very hot code*. The engine optimizes this code further. This optimization takes a long time (consider something like -O3-level optimization). This phase produces highly optimized code that runs super fast. This code is much faster than the previously optimized code and the interpreted version. Obviously, the engine spends more time during this phase, say in the order of milliseconds. This is compensated by the code performance and frequency of execution.

JavaScript is a dynamically typed language and all the optimizations the engine can do are based on the assumption of *types*. If the assumption breaks, then the code is interpreted and executed, and the optimized code gets removed rather than throwing a runtime exception. The JavaScript engine implements the necessary type checks and bails out the optimized code when the assumed type changes. But the time spent on the optimize phase is in vain.

We can prevent these *type*-related issues by using something such as **TypeScript**. TypeScript is a superset of JavaScript. With TypeScript, we can prevent polymorphic code (code that accepts different types). In the JavaScript engine, monomorphic code (code that accepts only one type) always runs faster than its polymorphic counterpart.

There is no use in having highly optimized monomorphic JavaScript code if the JavaScript files are huge in size. The JavaScript engine has to wait until the entire file is downloaded. With a poor connection, that takes forever to happen.

> **Note**
> It is important to split the JavaScript bundle into smaller chunks. Including the JavaScript asynchronously (or in other words, lazy loading) boosts the performance of your application. We need to strike a correct balance and know which JavaScript module/file to load, cache, and then revalidate. Larger file sizes (payloads) will degrade the performance of the application greatly.

The final step is **garbage collection**, where all the live objects in the memory are removed. The garbage collection in the JavaScript engine works on the basis of reference. During the garbage collection cycle, the JavaScript engine starts from the root object (something like global in Node.js). It finds all the objects referenced from the root object and marks them as reachable objects. It marks the remaining objects as unreachable objects. Finally, it sweeps the unreachable objects. Since it is automatically done by the JavaScript engine, the garbage collection process is not efficient and it is much slower.

Understanding WebAssembly execution inside the JavaScript engine

WASM is in binary format and is already compiled and optimized. The JavaScript engine fetches the WASM. Then, it decodes the WASM and converts it into the module's internal representation (that is, AST). This phase is called **decode**. The decode phase is much faster than JavaScript's **parse** phase.

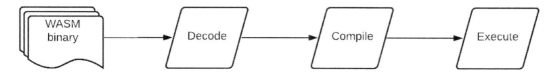

Figure 3.2 – WebAssembly execution inside the JavaScript engine

Next, the decoded WASM enters the **compile** phase. During this phase, the module is validated, and during the validation, the code is checked for certain conditions to guarantee the module is safe and does not have any harmful code. The functions, instruction sequences, and the usage of stacks are type-checked during the validation process. The validated code is then compiled to machine-executable code. Since the WASM is already compiled and optimized, this compile phase is faster. During this phase, the WASM is converted into machine code.

The compiled code then enters the **execute** phase. In the execute phase, the module is instantiated and invoked. During the instantiation, the engine instantiates the state and the execution stack (memory where it stores all the information related to the program) and then executes the module.

The other advantage of WebAssembly is that the module is ready to compile and instantiate right from the first byte. Thus, the JavaScript engine need not wait until the entire module is downloaded. This further increases WebAssembly's performance. WebAssembly is fast because its execution has fewer steps than JavaScript execution, so the binary is already optimized and compiled, and the binary can be streaming compiled.

> **Note**
>
> WASM does not always provide high performance. There are scenarios in which JavaScript performs better. So, it is necessary to understand that and think before using WebAssembly.

Find out more about JavaScript performance and how load time is involved at `https://medium.com/@addyosmani/the-cost-of-javascript-in-2018-7d8950fbb5d4`.

Find out more about chunking and code splitting in webpack at `https://webpack.js.org/guides/code-splitting/`.

We have seen how WebAssembly works inside the browser; now, let's explore the WebAssembly text format.

Exploring the WebAssembly text format

Machines understand a bunch of 1s and 0s. We optimize the binary to make it run faster and more efficiently. The more concise and optimal the instructions are, the more efficient and performant the machine will be. But for people, it is difficult to contextually analyze and understand a huge blob of 1s and 0s. That is the very reason why we started abstracting and creating high-level programming languages.

In the WebAssembly world, we convert human-readable programming languages, such as Rust, Go, and C/C++, into binary code. These binaries are a bunch of instructions with opcodes and operands. These instructions make the machine highly efficient but contextually make it difficult for us to understand.

Why should we worry about the readability of the binary generated? Because it helps us to understand the code, which helps while debugging the code.

WebAssembly provides the WebAssembly text format, WAST or WAT. WAST is a human-readable format of the WebAssembly binary. The JavaScript engine (both in the browser and Node.js), when loading the WebAssembly file, can convert the binary into WebAssembly text format. This helps in understanding what is in the code and debugging. Text editors can show the binary in WebAssembly text format, which is much more readable than its binary counterpart.

Basic WASM in binary format is as follows:

```
00 61 73 6d 01 00 00 00
```

This translates to the following:

```
00 61 73 6d 01 00 00 00
\0  a  s  m  1  0  0  0  (ascii value of the character)
|         |  |
---------  version
    |
Magic Header
```

This basic module has a magic header (\0asm) followed by the version of WebAssembly (01).

The textual format is written in an **s-expression format**. Every instruction/expression in s-expression syntax should live within a pair of parentheses, ( ). S-expressions are commonly used when defining a nested list or structured tree. Many research papers on tree-based data structures use this notation to showcase their code. The s-expression removes all the unnecessary ceremony from XML, providing a concise format.

> **Note**
>
> Does this expression (defining everything within parentheses) look familiar? Have you ever worked with LISP (or the languages that are built inspired by LISP)?

Modules are the basic building blocks in WASM. A textual representation of basic WASM is as follows:

```
(module )
```

WASM is made up of a header and zero or more sections. The header starts with a magic header and the version of WASM. Following the header, the WASM may have zero or more of the following sections:

- Types
- Functions
- Tables
- Memories
- Globals
- Element
- Data
- Start function
- Exports
- Imports

All these sections are optional in WASM. The structure of WASM looks as follows:

```
module ::= {
    types vec<funcType>,
    funcs vec<func>,
    tables vec<table>,
```

```
    mems vec<mem>,
    globals vec<global>,
    elem vec<elem>,
    data vec<data>,
    start start,
    imports vec<import>,
    exports vec<export>
}
```

Every section inside the WASM is a vector (array) that contains zero or more values of the respective types, except for `start`. We will explore the `start` section later in the book. For now, `start` holds an index that references a function in the `funcs` section.

Every section in the WASM takes the following format:

```
<section id><u32 section size><Actual content of the section>
```

The first byte refers to a unique section ID. Every section has a unique section ID. Next to the unique section ID is an **unsigned 32-bit (u32)** integer that defines the section's size in bytes. The remaining bytes are the section content.

> **Note**
> Since the section size is defined by a u32 integer, the maximum size of the section is limited to approximately 4.2 GB of memory (that is, 2^32 - 1).

In the WebAssembly text format, we use the name of the section to represent each segment in the section.

For example, the function section contains a list of functions. A sample function definition in WebAssembly text format is as follows:

```
(func <name>? <func_type> <local>* <inst>* )
```

As with other expressions, everything that we define goes within parentheses, `()`. First, we define the function block with a `func` keyword. Following the `func` keyword, we add the name of the function. The function name is optional here because in binary, the function is identified by the index of the function block inside the function section.

The name is followed by `func_type`. `func_type` is referred to as `type_use` in the spec. `type_use` here refers to the type definition. `func_type` holds all the input parameters (along with their types) and the return type of the function. So, for an `add` function, which takes two input operands and returns the result, `func_type` will look like this:

```
(param $lhs i32) (param $rhs i32) (result i32)
```

> **Note**
>
> The type is either `i32`, `i64`, `f32`, or `f64` (32-bit and 64-bit integer or float). The type information might change in the future, when WebAssembly adds support for more types.

The `param` keyword denotes the defined expression holds a parameter. `$lhs` is the variable name. Note that all variables defined in the WebAssembly text format will have `$` as a prefix. Following that, we have the type of the parameter, `i32`. Similarly, we have defined another expression for the second operand, `$rhs`. Finally, the return type is mentioned as `(result i32)`. The `result` keyword denotes that the expression is a return type, followed by the type, `i32`.

Following `func_type`, we define any local variables that we will use inside the function. Finally, we have a list of instructions/operations.

Let's define an `add` function with the preceding code snippets as a reference:

```
(func $add (param $lhs i32) (param $rhs i32) (result i32)
    get_local $lhs
    get_local $rhs
    i32.add)
```

The entire block is wrapped inside the parentheses. The function block starts with a `func` keyword. Then, we have an optional name (`$add`) for the function. The WebAssembly binary module will use the function index inside the function section to identify the function rather than a name. Then, we define the operand and return type.

> **Note**
>
> In binary format, the parameters and results are defined via the `type` section as that helps to optimize the generated functions. But in the text format, for brevity and ease of understanding, the type information will be shown in every function definition.

Then, we have a list of instructions. The first instruction, get_local, gets the local value of (from the heap) $lhs. Then, we fetch the local value of $rhs. After that, we add them both using the i32.add instruction. Finally, the closing parenthesis finishes things off.

There is no separate return statement/expression. So, how does the function know what to return?

As we have seen before, WebAssembly is a stack machine. When a function is called, it creates an empty stack for it. The function then uses this stack to push and pop data. So, when the get_local instruction is executed, it pushes the value into the stack. After the two get_local calls, the stack will have $lhs and $rhs in the stack. Finally, i32. add will pop two values from the stack, do the add operation, and push the element. When the function is ended, the top of the stack will be taken out and provided to the function caller.

If we want to export this function to the outside world, then we can add an export block:

```
(export <export_name> (func <function_reference>))
```

The export block is defined inside (). The export block starts with an export keyword. The export keyword is followed by the name of the function. Following the name, we refer to the function. The function block consists of the following func keyword. Then, we have function_reference, which refers to the name of the function defined/imported inside the module.

In order to export the add function, we define the following:

```
(export "add" (func $add))
```

"add" refers to the name with which the function is exported outside the module, followed by (func $add), referring to the function.

Both the function and export sections should be wrapped inside a module section, to make it valid WASM:

```
(module
    (func $add (param $lhs i32) (param $rhs i32)
      (result i32)
        get_local $lhs
        get_local $rhs
        i32.add)
    (export "add" (func $add))
)
```

The preceding is valid WASM. Imagine it as a tree structure with the module as its root and both the function and export as its children.

We have seen how to create a simple function in WebAssembly text format. Now, let's define a complex function in WebAssembly text format.

Building a function in WebAssembly text format

For this, we will use a recursive Fibonacci series generator. The Fibonacci function that we will be writing will be of the following format:

```
# Sample code in C for reference
int fib(n) {
    if (n <= 1)
        return 1;
    else
        return fib(n-1)+ fib(n-2);
}
```

Let's first define the function signature for the given `fib` function using WebAssembly text format. The `fib` function, similar to its C counterpart, takes in a number parameter and returns a number. So, the function definition follows the same signature in the WebAssembly text format:

```
(func $fib (param $n i32) (result i32)
    ...
)
```

We define the function inside parentheses, `()`. The function starts with a `func` keyword. Following the keyword, we add the function name, `$fib`. Then, we add the parameter to the function; in our case, the function has only one parameter, n; we define it as `(param $n i32)`. Then, the function returns a number, `(result i32)`.

WebAssembly does not have in-memory to handle temporary variables. In order to have local values, we should push the value into the stack and then retrieve it. So, to check n<=1, we have to first create a local variable and store 1 inside it, and then do the check. To define a local variable, we use the `local` block. The `local` block starts with a `local` keyword. This keyword is followed by the name of the variable. After the variable name, we define the type of the variable:

```
(local <name> <type>)
```

Let's create a `local` variable called $tmp:

```
(local $tmp i32)
```

> **Note**
>
> (local $tmp i32) is not an instruction. It is part of the function declaration. Remember, the preceding function syntax includes local.

We then have to set the value of $tmp to 1. To set the value, we first have to push the value 1 into the stack, after which we have to pop the value from the stack and set it to $tmp:

```
i32.const 1
set_local $tmp
```

i32.const creates an i32 constant value and pushes that into the stack. So, here, we create a constant with a value of 1 and push that into the stack.

Then, we set the value in $tmp using set_local. set_local takes the topmost value from the stack, in our case, 1, and assigns the value of $tmp to 1.

Now, we have to check whether the given parameter is less than 2. WebAssembly provides i32.<some_action> to do some action on i32. For example, to add two numbers, we have used i32.add. Similarly, to check whether it's less than a particular value, we have i32.lt_s. _s here denotes that we are checking for a signed number.

i32.lt_s expects two operands. For the first operand (that is, $n), we use the get_local expression to fetch the value from $n and put it at the top of the stack. Then, we create a constant of 2 using i32.const 2 and add 2 to the stack. Finally, we compare the $n value with 2 using i32.lt_s:

```
get_local $n
i32.const 2
i32.lt_s
```

But how do we define the *if condition*? WebAssembly provides br_if and block.

In WebAssembly text format, a block is defined with a block keyword followed by a name to identify the block. We end the block using end. The block looks as follows:

```
block $block
... ; some code goes in here.
end
```

We will provide this block to `br_if`. `br_if` calls the block if the condition succeeds:

```
get_local $n
i32.const 2
i32.lt_s
br_if $block ; calls the $block` only when the condition
   succeeds.
```

The WebAssembly text format so far will look like this:

```
(module
  (func $fib (param $n i32) (result i32) (local $tmp i32)
    i32.const 1
    set_local $tmp
    ; block
    block $block
      ; if condition
      get_local $n
      i32.const 2
      i32.lt_s
      br_if $block
    ... ; some code
    end
    ; return value
    get_local $tmp
  )
)
```

Everything is wrapped inside `module`. At the end of `$block`, the value will be stored in `$tmp`. We get the value of `$tmp` using `get_local $tmp`. The only thing that is left to do is to create the loop.

Loop time

First, we set `$tmp` to 1:

```
i32.const 1
set_local $tmp
```

Then, we will create a loop. To create a loop, the WebAssembly text format uses the loop keyword:

```
loop $loop
end
```

The loop keyword is followed by the name of the loop. The loop ends with the end keyword. loop is a special block that will run until we exit using some conditional expression such as br_if:

```
get_local $n
i32.const -2
i32.add
call $fib
get_local $tmp
i32.add
set_local $tmp
get_local $n
i32.const -1
i32.add
tee_local $n
i32.const 1
i32.gt_s
br_if $loop
```

We get $n and add -2 to it, and then call the fib function. To call a function, we use the call keyword followed by the name of the function. Here, call $fib returns the value and pushes the value into the stack.

Now, get $tmp using get_local $tmp. This pushes $tmp to the stack. Then, we use i32.add to pop two values from the stack and add them. Finally, we set $tmp using set_local $tmp. set_local $tmp takes the topmost value from the stack and assigns it to $tmp. We get $n and add -1 to it.

We use tee_local here because tee_local is similar to set_local but instead of pushing the value into the stack, it returns the value. Finally, we run the loop until $n is greater than 1. If it is less than 1, we break the loop using br_if $loop. The complete WebAssembly text format will look like this:

```
(module
  (func (export $fib (param $n i32) (result i32)
```

```
    (local $tmp i32)
    i32.const 1
    set_local $tmp
    ; block
    block $block
      ; if condition
      get_local $n
      i32.const 2
      i32.lt_s
      br_if $block
      ; loop
      loop $loop
        get_local $n
        i32.const -2
        i32.add
        call $fib
        get_local $tmp
        i32.add
        set_local $tmp
        get_local $n
        i32.const -1
        i32.add
        tee_local $n
        i32.const 1
        i32.gt_s
        br_if $loop
      end
    end
    ; return value
    get_local $tmp
  )
)
```

In future chapters, we will see how we can convert this WebAssembly text format into WASM and execute it.

If you're interested in learning more about s-expressions, check out `https://en.wikipedia.org/wiki/S-expression`.

To find out more about the WebAssembly text format design, check out the specifications at `https://github.com/WebAssembly/design/blob/master/Semantics.md`.

Check out more text instructions at `https://webassembly.github.io/spec/core/text/instructions.html`.

Refer to various instructions and their opcode at `https://webassembly.github.io/spec/core/binary/instructions.html`.

Find out more about binary encoding at `https://github.com/WebAssembly/design/blob/master/BinaryEncoding.md`.

Summary

In this chapter, we have seen how WebAssembly is executed inside the JavaScript engine and explored what WebAssembly text format is and how to define WASM using WebAssembly text format. In the next chapter, we will explore the WebAssembly Binary Toolkit.

Section 2: WebAssembly Tools

This section introduces the various tools available in WebAssembly and what we can achieve with them. It also explains various sections inside the WebAssembly module.

You will understand how to generate WebAssembly from different sources and how to use various tools in the WebAssembly ecosystem after finishing this chapter.

This section comprises the following chapters:

- *Chapter 4, Understanding WebAssembly Binary Toolkit*
- *Chapter 5, Understanding Sections in WebAssembly Module*
- *Chapter 6, Installing and Using Binaryen*

4
Understanding WebAssembly Binary Toolkit

The Rust compiler chain converts Rust code into WebAssembly binary. But the generated binaries are both size- and performance-optimized. It is difficult to understand, debug, and validate binary code (it is a bunch of hexadecimal numbers). Converting WebAssembly binary back into the original source code is very difficult. **WebAssembly Binary Toolkit (WABT)** helps to convert WebAssembly binary into a human-readable format, such as the **WebAssembly text (WAST)** format or C-native code.

> **Note**
> Native code here does not refer to the original source of truth; instead, it refers to C-native code that the machine interprets.

WebAssembly Binary Toolkit is abbreviated as WABT and pronounced as "*wabbit*." WABT provides a set of tools for converting, analyzing, and testing WebAssembly binaries.

In this chapter, we will explore WABT and how it helps to convert WebAssembly binary into various formats and why it is useful. We will cover the following main topics in this chapter:

- Getting started with WABT
- Converting WAST into WASM
- Converting WASM into WAST
- Converting WASM into C
- Converting WAST into JSON
- Understanding a few other tools provided by WABT

Technical requirements

You can find the code files present in this chapter on GitHub at `https://github.com/PacktPublishing/Practical-WebAssembly`.

Getting started with WABT

Let's first install WABT and then explore the various options provided by the WABT tool.

Installing WABT

In order to install WABT, first clone the repository from GitHub:

```
$ git clone --recursive https://github.com/WebAssembly/wabt
```

> **Note**
>
> We use the `--recursive` flag here as it ensures that after the clone is created, all submodules within the repository (such as `test-suite`) are initialized.

Go into the cloned repository, create a folder named `build`, and then go inside the `build` folder. This is where we will generate the binaries:

```
$ cd wabt
$ mkdir build
$ cd build
```

> **Note**
> You will also need to install CMake. Refer to `https://cmake.org/download/` for more instructions.

To build the binary with CMake, we first need to generate the build system. We specify the source to the `cmake` command. CMake will then build trees and generate a build system for the specified source, using the `CMakeLists.txt` file.

Linux or macOS

In order to generate the project build system, we run the `cmake` command with the path to the `wabt` folder. The `cmake` command accepts both a relative and absolute path. We are using the relative path here (`..`):

```
$ cmake ..
```

Now we can build the project using cmake build. `cmake build` makes use of the generated project binary tree to generate the binaries:

```
Usage: cmake --build <dir> [options] [-- [native-options]]
Options:
  <dir> = Project binary directory to be built.
  --parallel [<jobs>], -j [<jobs>]
        = Build in parallel using the given number of jobs.
                  If <jobs> is omitted the native build
                  tool's
                  default number is used.
                  The CMAKE_BUILD_PARALLEL_LEVEL
                  environment variable
                  specifies a default parallel level when
                  this option
                  is not given.
  --target <tgt>..., -t <tgt>...
                = Build <tgt> instead of default targets.
  --config <cfg> = For multi-configuration tools, choose
    <cfg>.
  --clean-first = Build target 'clean' first, then build.
                  (To clean only, use --target 'clean'.)
  --verbose, -v = Enable verbose output - if supported -
```

```
        including the build commands to be executed.
--                   = Pass remaining options to the native
    tool.
```

The cmake build command requires the <dir> option to generate the binaries. The cmake build command accepts the flags listed in the preceding code block:

```
$ cmake --build .
....
[100%] Built target spectest-interp-copy-to-bin
```

Windows

Install CMake and Visual Studio (>= 2015). Then, run cmake inside the build folder:

```
$ cmake [wabt project root] -DCMAKE_BUILD_TYPE=[config] -
    DCMAKE_INSTALL_PREFIX=[install directory] -G [generator]
```

The [config] parameter can be either DEBUG or RELEASE.

The [install directory] parameter should be the folder where you want to install the binaries.

The [generator] parameter should be the type of project you want to generate, for example, Visual Studio 14 2015. You can see the list of available generators by running cmake -help.

This will build and install all the required executables inside the folder specified:

```
$ cd build
$ cmake --build .. --config RELEASE --target install
```

Once you have successfully installed all the WABT tools, you can either add them to your path or call them from their path.

The build folder contains the following binaries:

```
$ tree -L 1
├── dummy
├── hexfloat_test
├── spectest-interp
├── wabt-unittests
├── wasm-c-api-global
```

```
├── wasm-c-api-hello
├── wasm-c-api-hostref
├── wasm-c-api-memory
├── wasm-c-api-multi
├── wasm-c-api-reflect
├── wasm-c-api-serialize
├── wasm-c-api-start
├── wasm-c-api-table
├── wasm-c-api-threads
├── wasm-c-api-trap
├── wasm-decompile
├── wasm-interp
├── wasm-objdump
├── wasm-opcodecnt
├── wasm-strip
├── wasm-validate
├── wasm2c
├── wasm2wat
├── wast2json
├── wat-desugar
└── wat2wasm
```

That sure is a huge list of binaries. Let's see what each one is capable of, in detail:

- wat2wasm – This tool helps to convert the WAST format into a **WebAssembly module (WASM)**.

- wat-desugar – This tool reads a file in a WASM S-expression and formats it.

- wast2json – This tool validates and converts WAST format into JSON format.

- wasm2wat – This tool converts WASM into WAST format.

- wasm2c – This tool converts WASM into C-native code.

- wasm-validate – This tool validates whether the given WebAssembly is constructed as per the specification.

- `wasm-strip` – As we saw in the previous chapter, WASM consists of various sections. The custom section in the module is used only for extra meta-information about the module and the tools used in its generation. `wasm-strip` removes the custom section from the WASM.

- `wasm-opcodecnt` – This tool reads the WASM and counts the use of opcode instructions in the WebAssembly Module.

- `wasm-objdump` – This tool helps to print information about a WASM binary. It is similar to objdump (`https://en.wikipedia.org/wiki/Objdump`) but for WebAssembly Modules.

- `wasm-interp` – This tool decodes and runs a WebAssembly binary file using a stack-based interpreter.

- `wasm-decompile` – This tool helps to decompile a WASM binary into readable C-like syntax.

The following are the testing binaries:

- `hexfloat_test`
- `spectest-interp`
- `wabt-unittests`

> **Note**
>
> Explore various supported proposals by WABT at `https://github.com/WebAssembly/wabt#supported-proposals`.

We have built WABT and generated the tools. Now, let's explore the most important and useful tools.

Converting WAST into WASM

`wat2wasm` helps to convert the WAST format into WASM. Let's take it for a spin:

1. Create a new folder called `wabt-playground` and go into the folder:

```
$ mkdir wabt-playground
$ cd wabt-playground
```

2. Create a `.wat` file called `add.wat`:

```
$ touch add.wat
```

3. Add the following contents to add.wat:

```
(module
(func $add (param $lhs i32) (param $rhs i32) (result i32)
 get_local $lhs
   get_local $rhs
     i32.add
 )
 )
```

4. Convert the WAST format into WASM using the wat2wasm binary:

```
$ /path/to/build/directory/of/wabt/wat2wasm add.wat
```

This generates a valid WebAssembly binary in add.wasm file:

```
00 61 73 6d 01 00 00 00 01 07 01 60 02 7f 7f 017f 03
  02 01 00 0a 09 01 07 00 20 00 20 01 6a 0b
```

Note that the size of the generated binary is 32 bytes.

5. WABT reads the WAST format file (.wat) and converts it into a WebAssembly module (.wasm). wat2wasm first validates the given file (.wat) and then converts it into a .wasm file. To check the various options supported by wat2wasm, we can run the following command:

```
$ /path/to/build/directory/of/wabt/wat2wasm --help
usage: wat2wasm [options] filename
  read a file in the wasm text format, check it for
  errors, and
  convert it to the wasm binary format.
examples:
  # parse and typecheck test.wat
  $ wat2wasm test.wat
  # parse test.wat and write to binary file test.wasm
  $ wat2wasm test.wat -o test.wasm
  # parse spec-test.wast, and write verbose output to
    stdout (including
  # the meaning of every byte)
  $ wat2wasm spec-test.wast -v
options:
```

```
        --help              Print this help message
        --version           Print version information
   -v, --verbose            Use multiple times for more info
        --debug-parser      Turn on debugging the parser
                            of wat files
   ...
        --debug-names       Write debug names to the
                            generated binary file
        --no-check          Don't check for invalid
                            modules
```

If we need to generate the WASM file in a different name, we can use the -o option with the filename. For example, wat2wasm add.wat -o add.wasm will generate add.wasm.

6. wat2wasm also provides verbose output that clearly explains how the WASM is structured. In order to see the structure of the WASM, we run it with the -v option:

```
$ /path/to/build/directory/of/wabt/wat2wasm add.wat -v
0000000: 0061 736d          ; WASM_BINARY_MAGIC
0000004: 0100 0000          ; WASM_BINARY_VERSION
; section "Type" (1)
0000008: 01                 ; section code
0000009: 00                 ; section size (guess)
000000a: 01                 ; num types
; type 0
000000b: 60                 ; func
000000c: 02                 ; num params
000000d: 7f                 ; i32
000000e: 7f                 ; i32
000000f: 01                 ; num results
0000010: 7f                 ; i32
0000009: 07                 ; FIXUP section size
; section "Function" (3)
0000011: 03                 ; section code
0000012: 00                 ; section size (guess)
0000013: 01                 ; num functions
```

```
0000014: 00              ; function 0 signature index
0000012: 02              ; FIXUP section size
; section "Code" (10)
0000015: 0a              ; section code
0000016: 00              ; section size (guess)
0000017: 01              ; num functions
; function body 0
0000018: 00              ; func body size (guess)
0000019: 00              ; local decl count
000001a: 20              ; local.get
000001b: 00              ; local index
000001c: 20              ; local.get
000001d: 01              ; local index
000001e: 6a              ; i32.add
000001f: 0b              ; end
0000018: 07              ; FIXUP func body size
0000016: 09              ; FIXUP section size
```

The preceding output is a detailed description of the binary generated. The leftmost seven numbers are the index, followed by a colon. The next two characters are the actual binary code, and then comments. The comment describes what the binary (op)code does.

The first two lines specify wasm_magic_header and its version. The next segment is the Type section. The Type section defines the section ID, followed by the size of the section, and then the number of type blocks. In our case, we have only one type. The type 0 section defines the type signature of the add function.

Then, we have the Function section. In the Function section, we have the section ID, followed by the section size, and then the number of functions. The function section does not have the function body. The function body is defined in the Code section.

While generating the binary, we can enable the compiler to include the new and shiny features and disable various existing features using the appropriate enable-* and disable-* options, respectively.

> **Note**
>
> You can also check the online version at https://webassembly.github.io/wabt/demo/wat2wasm/ to explore the WABT tools.

We have converted WAST into WASM. Now, let's explore how to convert WASM into WAST with wasm2wat.

Converting WASM into WAST

Sometimes, for debugging or understanding, we need to know what the WASM is doing. WABT has a wasm2wat converter. It helps to convert WASM into WAST format:

```
$ /path/to/build/directory/of/wabt/wasm2wat add.wasm
(module
  (type (;0;) (func (param i32 i32) (result i32)))
  (func (;0;) (type 0) (param i32 i32) (result i32)
    local.get 0
    local.get 1
    i32.add))
```

Running the previous command will convert add.wasm back into WAST format and print the output in the console.

If you want to save it as a file, you can do so by using the -o flag:

```
$ /path/to/build/directory/of/wabt/wasm2wat add.wasm -o new_
  add.wat
```

This command creates a new_add.wat file.

To check the various options supported by wasm2wat, we can run the following command:

```
$ wasm2wat --help
usage: wasm2wat [options] filename

  Read a file in the WebAssembly binary format, and convert
  it to
  the WebAssembly text format.

examples:
  # parse binary file test.wasm and write text file test.wast
  $ wasm2wat test.wasm -o test.wat
```

```
   # parse test.wasm, write test.wat, but ignore the debug
names, if any
   $ wasm2wat test.wasm --no-debug-names -o test.wat


options:
         --help                          Print this help
           message
         --version                       Print version
           information
    -v, --verbose                        Use multiple times
      for more info
    -o, --output=FILENAME                Output file for the
      generated wast file, by default use stdout
    -f, --fold-exprs                     Write folded
      expressions where possible
    ....
         --no-debug-names                Ignore debug names
           in the binary file
         --ignore-custom-section-errors  Ignore errors in
           custom sections
         --generate-names                Give auto-generated
           names to non-named functions, types, etc.
         --no-check                      Don't check for
           invalid modules
```

> **Note**
>
> Both wasm2wat and wat2wasm have almost identical options.

Running the previous command with the -v option prints the AST syntax of the
WAST format:

```
$ /path/to/build/directory/of/wabt/wasm2wat add.wasm -o new_
   add.wat -v
BeginModule(version: 1)
   BeginTypeSection(7)
     OnTypeCount(1)
```

```
    OnType(index: 0, params: [i32, i32], results: [i32])
  EndTypeSection
  BeginFunctionSection(2)
    OnFunctionCount(1)
    OnFunction(index: 0, sig_index: 0)
  EndFunctionSection
  BeginCodeSection(9)
    OnFunctionBodyCount(1)
    BeginFunctionBody(0, size:7)
    OnLocalDeclCount(0)
    OnLocalGetExpr(index: 0)
    OnLocalGetExpr(index: 1)
    OnBinaryExpr("i32.add" (106))
    EndFunctionBody(0)
  EndCodeSection
  BeginCustomSection('name', size: 28)
    BeginNamesSection(28)
      OnNameSubsection(index: 0, type: function, size:6)
      OnFunctionNameSubsection(index:0, nametype:1, size:6)
      OnFunctionNamesCount(1)
      OnFunctionName(index: 0, name: "add")
      OnNameSubsection(index: 1, type: local, size:13)
      OnLocalNameSubsection(index:1, nametype:2, size:13)
      OnLocalNameFunctionCount(1)
      OnLocalNameLocalCount(index: 0, count: 2)
      OnLocalName(func_index: 0, local_index: 0, name: "lhs")
      OnLocalName(func_index: 0, local_index: 1, name: "rhs")
    EndNamesSection
  EndCustomSection
EndModule
```

The entire code block is wrapped inside BeginModule and EndModule.
BeginModule includes the version of the WebAssembly binary.

Inside the BeginModule, the BeginTypeSection starts with the section index of
type (that is, 7), followed by OnTypeCount, the number of types defined. Then, we
have the actual definition of the type with OnType. We end the type section with
EndTypeSection.

Then, we have the Function section marked by BeginFunctionSection and EndFunctionSection. This contains the function count (OnFunctionCount) and the function definition (OnFunction).

Finally, we have the code section, which holds the actual body of the function. The code section begins and ends with BeginCodeSection and EndCodeSection.

Sometimes, WASM may contain debug names. We can ignore them using the --no-debug-names flag:

```
$ /path/to/build/directory/of/wabt/wasm2wat add.wasm -o new_
  add.wat -v --no-debug-names
BeginModule(version: 1)
  BeginTypeSection(7)
    OnTypeCount(1)
    OnType(index: 0, params: [i32, i32], results: [i32])
  EndTypeSection
  BeginFunctionSection(2)
    OnFunctionCount(1)
    OnFunction(index: 0, sig_index: 0)
  EndFunctionSection
  BeginCodeSection(9)
    OnFunctionBodyCount(1)
    BeginFunctionBody(0, size:7)
    OnLocalDeclCount(0)
    OnLocalGetExpr(index: 0)
    OnLocalGetExpr(index: 1)
    OnBinaryExpr("i32.add" (106))
    EndFunctionBody(0)
  EndCodeSection
  BeginCustomSection('name', size: 28)
  EndCustomSection
EndModule
```

Note BeginCustomSection and EndCustomSection. Compare it with the previous output; it is missing NamesSection. Now, let's check out the various options provided by the wasm2wat tool.

-f or --fold-exprs

As a big fan of functional programming, this is one of the coolest options available. It folds over the expression; that is, it converts expression 1 >> expression 2 >> operation into operation >> expression 1 >> expression2.

Let's see that in action:

1. Create a WAST file called `fold.wat` and fill it with the following contents:

```
(module
    (func $fold (result i32)
        i32.const 22
        i32.const 20
        i32.add
    )
)
```

2. Let's first convert it into WASM using `wat2wasm`:

```
$ /path/to/build/directory/of/wabt/wat2wasm -v
  fold.wat
; some contents
0000018: 41                                              ;
  i32.const
0000019: 16                                              ;
  i32 literal
000001a: 41                                              ;
  i32.const
000001b: 14                                              ;
  i32 literal
000001c: 6a                                              ;
  i32.add
; other contents
```

This creates `fold.wasm`.

3. Now, convert the WASM into the WAST format using `wasm2wat` and pass in the -f option:

```
$ /path/to/build/directory/of/wabt/wasm2wat -v
  fold.wasm -o converted_fold.wat -f
```

This will create a file called `converted_fold.wat`:

```
(module
    (type (;0;) (func (result i32)))
    (func (;0;) (type 0) (result i32)
        (i32.add
            (i32.const 1)
            (i32.const 2))))
```

Instead of using `i32.const 1` (expression 1) and `i32.const 2` (expression 2) and then doing `i32.add` (operation), this generates an output, `i32.add` (operation), followed by `i32.const 1` (expression 1) and `i32.const 2` (expression 2).

While generating `wat`, we can enable the compiler to include the new and shiny features and disable various existing features using the appropriate `enable-*` and `disable-*` options, respectively.

We have converted WASM into WAST. Now, let's explore how to convert WASM into native code (C) using `wasm2c`.

Converting WASM into C

WABT has a `wasm2c` converter that converts WASM into C source code and a header.

Let's create a `simple.wat` file:

```
$ touch simple.wat
```

Add the following contents to `simple.wat`:

```
(module
    (func $uanswer (result i32)
        i32.const 22
        i32.const 20
        i32.add
    )
)
```

`wat` here defines a `uanswer` function that adds `22` and `20` to give `42` as the answer. Let's create a WebAssembly binary using `wat2wasm`:

```
$ /path/to/build/directory/of/wabt/wat2wasm simple.wat -o
  simple.wasm
```

This generates the `simple.wasm` binary. Now, convert the binary into C code using `wasm2c`:

```
$ /path/to/build/directory/of/wabt/wasm2c simple.wasm -o
  simple.c
```

This generates `simple.c` and `simple.h`.

> **Note**
>
> Both `simple.c` and `simple.h` might look huge. Remember this is an autogenerated file and it includes all the necessary headers and configuration needed for the program to run.

simple.h

`simple.h` (the header file) includes standard boilerplate for the header. It also includes the `_cplusplus` condition to prevent name mangling in C++:

```
#ifndef SIMPLE_H_GENERATED_
#define SIMPLE_H_GENERATED_
#ifdef __cplusplus
extern "C" {
#endif

...

#ifdef __cplusplus
}
#endif
#endif
```

Since we have used `i32.const` and `i32.add`, the header file also imports `stdint.h`. It includes `wasm-rt.h`. The `wasm-rt.h` header imports the necessary WASM runtime variables.

Next, we can specify a module prefix. The module prefix is useful when using multiple modules. Since we only have one module, we use an empty prefix:

```
#ifndef WASM_RT_MODULE_PREFIX
#define WASM_RT_MODULE_PREFIX
#endif

#define WASM_RT_PASTE_(x, y) x ## y
#define WASM_RT_PASTE(x, y) WASM_RT_PASTE_(x, y)
#define WASM_RT_ADD_PREFIX(x) WASM_RT_PASTE(WASM_RT_MODULE_
PREFIX, x)
```

Next, we have some typedefs for the various number formats that WASM supports:

```
typedef uint8_t u8;
typedef int8_t s8;
typedef uint16_t u16;
typedef int16_t s16;
typedef uint32_t u32;
typedef int32_t s32;
typedef uint64_t u64;
typedef int64_t s64;
typedef float f32;
typedef double f64;
```

simple.c

`simple.c` provides the actual C code generated from the WASM binary. The generated code has the following:

1. We'll require the following list of libraries to use in the code:

    ```
    #include <math.h>
    #include <string.h>

    #include "simple.h"
    ```

2. Next, we define the trap that is called when an error occurs:

```
#define TRAP(x) (wasm_rt_trap(WASM_RT_TRAP_##x), 0)
```

3. Then, we define PROLOGUE, EPILOGUE, and UNREACHABLE_TRAP, which are called before the start of execution, after execution, and when the execution meets an unreachable exception, respectively:

```
#define FUNC_PROLOGUE
\
  if (++wasm_rt_call_stack_depth >
    WASM_RT_MAX_CALL_STACK_DEPTH) \
    TRAP(EXHAUSTION)


#define FUNC_EPILOGUE --wasm_rt_call_stack_depth

#define UNREACHABLE TRAP(UNREACHABLE)
```

WASM_RT_MAX_CALL_STACK_DEPTH is the maximum depth of the stack. By default, the value is 500 but we can change it. Note that if it reaches the limit, then an exception is thrown.

4. Next, we define the memory manipulations:

```
#if WASM_RT_MEMCHECK_SIGNAL_HANDLER
#define MEMCHECK(mem, a, t)
#else
#define MEMCHECK(mem, a, t)   \
  if (UNLIKELY((a) + sizeof(t) > mem->size)) TRAP(OOB)
#endif

#define DEFINE_LOAD(name, t1, t2, t3)
\
  static inline t3 name(wasm_rt_memory_t* mem, u64
    addr) {   \
    MEMCHECK(mem, addr, t1);
\
    t1 result;
\
    __builtin_memcpy(&result, &mem->data[addr],
```

```
        sizeof(t1)); \
    return (t3)(t2)result;
  \
  }


#define DEFINE_STORE(name, t1, t2)
  \
  static inline void name(wasm_rt_memory_t* mem, u64
    addr, t2 value) { \
    MEMCHECK(mem, addr, t1);
  \
    t1 wrapped = (t1)value;
  \
    __builtin_memcpy(&mem->data[addr], &wrapped,
      sizeof(t1));              \
  }
```

MEMCHECK checks for memory. The DEFINE_LOAD and DEFINE_STORE blocks
define how to load and store a value in memory.

5. Next, we define a bunch of load and store operations for various data types that we
    have in this example:

```
DEFINE_LOAD(i32_load, u32, u32, u32);
DEFINE_LOAD(i64_load, u64, u64, u64);
...
DEFINE_LOAD(i64_load32_u, u32, u64, u64);
DEFINE_STORE(i32_store, u32, u32);
DEFINE_STORE(i64_store, u64, u64);
...
DEFINE_STORE(i64_store32, u32, u64);
```

Then, we define various functions that each of the data types supports, such as
TRUNC, DIV, and REM.

6. Next, we initialize the function types using func_types:

```
static u32 func_types[1];

static void init_func_types(void) {
```

```
func_types[0] = wasm_rt_register_func_type(0, 1,
    WASM_RT_I32);
}
```

This registers the (result i32) type in the WASM provided.

7. Next, we initialize `globals`, `memory`, `table`, and `exports`:

```
static void init_globals(void) {
}

static void init_memory(void) {
}

static void init_table(void) {
  uint32_t offset;
}

static void init_exports(void) {
}
```

8. Now, we implement the actual function:

```
static u32 w2c_f0(void) {
  FUNC_PROLOGUE;
  u32 w2c_i0, w2c_i1;
  w2c_i0 = 22u;
  w2c_i1 = 20u;
  w2c_i0 += w2c_i1;
  FUNC_EPILOGUE;
  return w2c_i0;
}
```

The function is static. It calls FUNC_PROLOGUE before the execution. Then, it creates two variables (both are unsigned u32). Then, we define the value of both the variables, 22 and 20, respectively. After that, we add them both. Once the execution is complete, we call FUNC_EPILOGUE. Finally, we return the value.

> **Note**
> Since we have not exported anything in wat, init_exports is empty.

We have converted WASM into C. The generated code is slightly different from the original code. Let's explore how to convert WAST into JSON with `wast2json`.

Converting WAST into JSON

The `wast2json` tool reads the WAST format and parses it, checks for errors, and then converts WAST into the JSON file. It generates a JSON and WASM file associated with the WAST file. Then, it links the WASM inside the JSON:

```
$ /path/to/build/directory/of/wabt/wast2json add.wat -o add.
  json
$ cat add.json
{"source_filename": "add.wat",
"commands": [
{"type": "module", "line": 1, "filename": "add.0.wasm"}]}
```

To check the various options supported by `wast2json`, run the following command:

```
$ /path/to/build/directory/of/wabt/wast2json --help
usage: wast2json [options] filename

   read a file in the wasm spec test format, check it for
   errors, and
   convert it to a JSON file and associated wasm binary files.

examples:
   # parse spec-test.wast, and write files to spec-test.json.
   Modules are
   # written to spec-test.0.wasm, spec-test.1.wasm, etc.
   $ wast2json spec-test.wast -o spec-test.json

options:
      --help                     Print this help
         message
      --version                  Print version
         information
    -v, --verbose                Use multiple times
      for more info
```

```
    --debug-parser                          Turn on debugging
      the parser of wast files
    --enable-exceptions                     Enable Experimental
      exception handling
    --disable-mutable-globals               Disable Import/
      export mutable globals
    --disable-saturating-float-to-int       Disable Saturating
      float-to-int operators
    --disable-sign-extension                Disable Sign-
      extension operators
    --enable-simd                           Enable SIMD support
    --enable-threads                        Enable Threading
      support
    --disable-multi-value                   Disable Multi-value
    --enable-tail-call                      Enable Tail-call
      support
    --enable-bulk-memory                    Enable Bulk-memory
      operations
    --enable-reference-types                Enable Reference
      types (externref)
    --enable-annotations                    Enable Custom
      annotation syntax
    --enable-gc                             Enable Garbage
      collection
    --enable-memory64                       Enable 64-bit
      memory
    --enable-all                            Enable all features
-o, --output=FILE                           output JSON file
-r, --relocatable                           Create a
  relocatable wasm binary (suitable for linking with e.g. lld)
    --no-canonicalize-leb128s               Write all LEB128
      sizes as 5-bytes instead of their minimal size
    --debug-names                           Write debug names
      to the generated binary file
    --no-check                              Don't check for
      invalid modules
```

These are the frequently used WABT tools. There are a few other tools provided by WABT that help with debugging and understanding WASM better.

Understanding a few other tools provided by WABT

In addition to the converters, WABT also provides a few tools that help us to understand WASM better. In this section, let's explore the following tools provided by WABT:

- `wasm-objdump`
- `wasm-strip`
- `wasm-validate`
- `wasm-interp`

wasm-objdump

Object code is nothing more than a sequence of instructions or statements in the computer language. Object code is what the compiler produces. The object code is then collected together and then stored inside the object file. The object file is the metadata holder for linking and debugging information. The machine code in the object file is not directly executable, but it provides valuable information when debugging and also helps with linking.

> **Note**
> `objdump` is the tool that is available in POSIX systems that provides a way to disassemble the binary format and print the assembly format of the code that is running.

The `wasm-objdump` tool provides the following options:

```
$ /path/to/build/directory/of/wabt/wasm-objdump --help
usage: wasm-objdump [options] filename+

  Print information about the contents of wasm binaries.

examples:
  $ wasm-objdump test.wasm
```

```
options:
        --help                  Print this help message
        --version               Print version information
  -h,   --headers               Print headers
  -j,   --section=SECTION       Select just one section
  -s,   --full-contents         Print raw section contents
  -d,   --disassemble           Disassemble function bodies
        --debug                 Print extra debug information
  -x,   --details               Show section details
  -r,   --reloc                 Show relocations inline with
    disassembly
```

At least one of the following options should be provided to the wasm-objdump command:

```
-d/--disassemble
-h/--headers
-x/--details
-s/--full-contents
```

The -h option prints all the available headers in WASM. For example, in our add example (add.wasm), we have the following:

```
$ /path/to/build/directory/of/wabt/wasm-objdump add.wasm -h
add.wasm: file format wasm 0x1

Sections:

     Type start=0x0000000a end=0x00000011 (size=0x00000007)
  count: 1
Function start=0x00000013 end=0x00000015 (size=0x00000002)
  count: 1
     Code start=0x00000017 end=0x00000020 (size=0x00000009)
  count: 1
```

Here, we have three sections available:

- Type
- Function

- Code

The -d option prints the actual body of the function available:

```
$/path/to/build/directory/of/wabt/wasm-objdump add.wasm -d
add.wasm: file format wasm 0x1


Code Disassembly:

000019 func[0]:
00001a: 20 00 | local.get 0
00001c: 20 01 | local.get 1
00001e: 6a    | i32.add
00001f: 0b    | end
```

It dissembles the assembly function and prints only the function body.

The -x option prints the section details of the WebAssembly binary file:

```
$ /path/to/build/directory/of/wabt/wasm-objdump add.wasm -x

add.wasm: file format wasm 0x1

Section Details:

Type[1]:
- type[0] (i32, i32) -> i32
Function[1]:
- func[0] sig=0
Code[1]:
- func[0] size=7
```

The -s option prints the contents of all the sections that are available:

```
$ /path/to/build/directory/of/wabt/wasm-objdump add.wasm -s

add.wasm: file format wasm 0x1
```

```
Contents of section Type:
000000a: 0160 027f 7f01 7f  .`.....

Contents of section Function:
0000013: 0100  ..

Contents of section Code:
0000017: 0107 0020 0020 016a 0b  ... . .j.
```

wasm-strip

The custom section in a WASM holds information about the names of the function and all locals defined in the WASM. It may contain information about the build and how the WASM was created. This is additional information. It bloats the binary and does not add any functionality.

We can strip the custom section to trim the binary size using the wasm-strip tool:

1. Create a wat file with the following contents:

    ```
    $ touch simple.wat
    (module
        (func $fold (result i32)
            i32.const 22
            i32.const 20
            i32.add
        )
    )
    ```

2. Now, convert that into a WASM with wat2wasm:

    ```
    $ /path/to/build/directory/of/wabt/wat2wasm simple.wat
      --debug-names
    $ l simple.wasm
    51B simple.wasm
    ```

> **Note**
>
> The --debug-names option provided generates the custom section and adds it to the binary generated.

The previous command generates a simple.wasm file and it is 51 bytes in size.

3. Now, let's remove the custom section from the binary using the following:

```
$ /path/to/build/directory/of/wabt/wasm-strip add.wasm
$ l simple.wasm
30B simple.wasm
```

As you can see, it removed 21 bytes of unnecessary information. Some WASM generators add the custom section for a better debugging experience but when deploying in production, we do not need the custom section. Use wasm-strip to remove it.

wasm-validate

We can validate the WASM using wasm-validate:

1. Create error.wasm with the following contents:

```
00 61 73 6d 03 00 00 00
            |
        Error value
```

2. Now, run wasm-validate to check whether the WASM is valid or not:

```
$ /path/to/build/directory/of/wabt/wasm-validate
  error.wasm
0000004: error: bad magic value
```

3. The wasm-validate tool provides the following options:

```
usage: wasm-validate [options] filename
```

4. Read a file in the WebAssembly binary format and validate it:

```
examples:
  # validate binary file test.wasm
  $ wasm-validate test.wasm


options:
```

```
    --help                    Print this help message
    --version                 Print version information
-v, --verbose                 Use multiple times for
  more info
    --enable-exceptions    Enable Experimental
      exception handling
    --disable-mutable-globals        Disable
      Import/export mutable globals
    --disable-saturating-float-to-int          Disable
      Saturating float-to-int operators
    --disable-sign-extension                Disable
      Sign-extension operators
    --enable-simd                       Enable
      SIMD support
    --enable-threads                    Enable
      Threading support
    --disable-multi-value               Disable
      Multi-value
    --enable-tail-call                  Enable
      Tail-call support
    --enable-bulk-memory                Enable
      Bulk-memory operations
    --enable-reference-types            Enable
      Reference types (externref)
    --enable-annotations                Enable
      Custom annotation syntax
    --enable-gc                         Enable
      Garbage collection
    --enable-memory64                   Enable
      64-bit memory
    --enable-all                        Enable
      all features
    --no-debug-names                    Ignore
      debug names in the binary file
    --ignore-custom-section-errors          Ignore
      errors in custom sections
```

wasm-interp

wasm-interp reads a file in the WebAssembly binary format and runs it in a stack-based interpreter. The wasm-interp tool parses the binary file and then type checks it:

```
$ /path/to/build/directory/of/wabt/wasm-interp add.wasm -v
BeginModule(version: 1)
  BeginTypeSection(7)
    OnTypeCount(1)
    OnType(index: 0, params: [i32, i32], results: [i32])
  EndTypeSection
  BeginFunctionSection(2)
    OnFunctionCount(1)
    OnFunction(index: 0, sig_index: 0)
  EndFunctionSection
  BeginCodeSection(9)
    OnFunctionBodyCount(1)
    BeginFunctionBody(0, size:7)
    OnLocalDeclCount(0)
    OnLocalGetExpr(index: 0)
    OnLocalGetExpr(index: 1)
    OnBinaryExpr("i32.add" (106))
    EndFunctionBody(0)
  EndCodeSection
EndModule
    0| local.get $2
    8| local.get $2
   16| i32.add %[-2], %[-1]
   20| drop_keep $2 $1
   32| return
```

The last five lines are how the stack interpreter executes the code.

The wasm-interp tool provides the following options:

```
usage: wasm-interp [options] filename [arg]...

  read a file in the wasm binary format and run it in a stack-
```

```
based
interpreter.

examples:
    ...

options:
        --help                          Print this help
            message
        --version                       Print version
            information
    ...
```

WABT provides a list of tools that make WASM easier to understand, debug, and convert into various human-readable formats. It is one of the most important toolkits that allows developers to explore the WASM better.

Summary

In this chapter, we saw what WABT is and how to install and use various tools provided by it. The WABT tool is very important in the WebAssembly ecosystem as it provides an easy option to convert non-readable, compact binaries into readable, expanded source code.

In the next chapter, we will explore various sections inside the WASM.

5
Understanding Sections in WebAssembly Modules

A WebAssembly module is composed of zero or more sections. Each section has its own functionality. In the previous chapters, we saw how functions are defined inside a WebAssembly module. A function is a section inside a WebAssembly module.

In this chapter, we will explore the various other sections inside a WebAssembly module. Understanding the various sections inside a WebAssembly module will make it easier for us to identify, debug, and write efficient WebAssembly modules. We will cover the following sections in this chapter:

- Exports and imports
- Globals
- Start
- Memory

Technical requirements

You can find the code files present in this chapter on GitHub at `https://github.com/PacktPublishing/Practical-WebAssembly/tree/main/05-wasm-sections`.

Exports and imports

A WebAssembly module consists of export and import sections. These sections are responsible for exporting functions out of and importing functions into the WebAssembly module.

Exports

In order to call the functions defined in a WebAssembly module from JavaScript, we need to export the functions from the WebAssembly module. The export section is where we will define all the functions that are exported out of the WebAssembly module.

Let's go back to our classic `add.wat` example from the previous chapter:

```
; add.wat
(module
    (func $add (param $lhs i32) (param $rhs i32)
      (result i32)
        get_local $lhs
        get_local $rhs
        i32.add)
    (export "add" (func $add))
)
```

Here, we have exported the add function using the `(export "add" (func $add))` statement. To export a function, we have used the `export` keyword followed by the name of the function and then the pointer to the exported function itself.

Remember that WebAssembly is compact-sized. Thus, we can represent the export statement along with the function definition itself, like so:

```
; add.wat
(module
    (func $add (export "add") (param $lhs i32)
      (param $rhs i32) (result i32)
```

```
        get_local $lhs
        get_local $rhs
        i32.add)
)
```

Let's use WABT's `wat2wasm` tool to convert the WebAssembly text format into a WebAssembly module with the following command:

$ /path/to/wabt/bin/wat2wasm add.wat

Let's analyze the generated byte code using the `hexdump` tool:

```
$ hexdump add.wasm
0000000 00 61 73 6d 01 00 00 00 01 07 01 60 02 7f 7f 01
0000010 7f 03 02 01 00 07 07 01 03 61 64 64 00 00 0a 09
0000020 01 07 00 20 00 20 01 6a 0b
0000029
```

As expected, the first byte consists of the magic header and version of the binary `00 61 73 6d 01 00 00 00`:

```
0000000: 0061 736d                    ; WASM_BINARY_MAGIC
0000004: 0100 0000                    ; WASM_BINARY_VERSION
```

The next bit is `01`, which represents the section index of the type section. Following that, we have the size of the type section, which is `07`. The next seven bits are the type section. `01` represents the number of type definitions available:

```
; section "Type" (1)
0000008: 01                           ; section code
0000009: 07                           ; section size
000000a: 01                           ; num types
```

Then, we have `60`, which represents `func`. Following that, we have `02`, representing the two parameters. `7f` is the opcode for defining the i32 type. Since both the parameters are i32, we have consecutive `7f` opcodes. Following that, the last two bits represent the return type and there is also `7f` representing i32:

```
; type 0
000000b: 60                           ; func
000000c: 02                           ; num params
```

```
000000d: 7f                                    ; i32
000000e: 7f                                    ; i32
000000f: 01                                    ; num results
0000010: 7f                                    ; i32
```

After the `type` section, we have the `func` section. The unique identifier for the `func` section is `03`. Following that, we have `02`, which defines the size of the function section. That is the size of the function section is just 2 bits. But we defined the function definition for `add` in the WebAssembly text format and the function is more than 2 bits in size. So, how is it possible? The reason is that the function section does not have the body of the function; instead, it just defines the available functions. The functions are defined in the code section. The next `01` defines that there is only one function defined in the module:

```
; section "Function" (3)
0000011: 03                                    ; section code
0000012: 02                                    ; section size
0000013: 01                                    ; num functions
0000014: 00                                    ; function 0 signature
  index
```

Then, we have the export section, which starts with `07`. The next `07` represents the size of the export section. Then, we define the number of exports exported in the export section. The next bit represents the length of the exported function name. The next `03` bits represent the function name, `add`. Then, the export section has the kind of export and function index of the exported function:

```
; section "Export" (7)
0000015: 07                                    ; section code
0000016: 07                                    ; section size
0000017: 01                                    ; num exports
0000018: 03                                    ; string length
0000019: 6164 64                    add        ; export name
000001c: 00                                    ; export kind
000001d: 00                                    ; export func index
```

The last segment starts with `0a`. `0a` is a unique identifier for the code section. The code section is of length `09`. Next, `01` represents the number of functions defined in the code block.

Next, 07 represents the length of the function definition. The next seven bits actually define the function block. 00 indicates that the function block does not have any local declarations. 20 is the opcode for get_local and we take the 00 index, and then again we have 20 opcode to get_local and we take the 01 index. Then, we add them using i32.add. The opcode for the i32 addition is 6a. Finally, we use 0b to end the function code block:

```
; section "Code" (10)
000001e: 0a                              ; section code
000001f: 09                              ; section size
0000020: 01                              ; num functions
  ; function body 0
0000021: 07                              ; func body size

0000022: 00                              ; local decl count
0000023: 20                              ; local.get
0000024: 00                              ; local index
0000025: 20                              ; local.get
0000026: 01                              ; local index
0000027: 6a                              ; i32.add
0000028: 0b                              ; end
```

We have seen how the export section is represented in a WebAssembly module. In the next section, let's see how the import section is represented in a WebAssembly module.

Imports

In order to import a function from another WebAssembly module or JavaScript module, we need to import the functions in the WebAssembly module. The import section is where we will import all the external dependencies into the WebAssembly module.

Now, let's imagine that a JavaScript module exports a function named jsAdd. We can import the jsAdd function using the import keyword. Create a file called jsAdd.wat and add the following content to it:

```
(module
    (func $i (import "imports" "jsAdd") (param i32))
)
```

Here, we are defining a function with the `func` keyword, followed by the name of the function, `$i`. We use `$i` to call the function inside the WebAssembly module. Then, we have the `import` keyword. The `import` keyword is followed by the module name. The module name here refers to the JavaScript module, and then we have the name of the function to import from the JavaScript module.

Finally, we have `param`. Since a WebAssembly module is typed, we have to define the input parameters and return types in the function definition.

Let's use WABT's `wat2wasm` to convert the WebAssembly text format into a WebAssembly module with the following command:

```
$ /path/to/wabt/bin/wat2wasm jsAdd.wat
```

Let's analyze the generated byte code using the `hexdump` tool:

```
$ hexdump jsAdd.wasm
0000000 00 61 73 6d 01 00 00 00 01 05 01 60 01 7f 00 02
0000010 11 01 07 69 6d 70 6f 72 74 73 05 6a 73 41 64 64
0000020 00 00
0000022
```

The binary consists of the import section, which starts at the index of 16. The import section starts with `02` because the unique section index of the import section is `02`. After that, we have `11`, which represents the size of the import section in the binary. The next bit represents the number of imports, `01`.

Then, we have the definition for the import. `07` here represents the length of the imported function. The next seven bits represent the name of the import module. The next bit represents the length of the function name, `05`, and the next five bits represent the function name. Finally, we have the kind and type signature of the index:

```
; Other information
; section "Import" (2)
000000f: 02                              ; section code
0000010: 11                              ; section size
0000011: 01                              ; num imports
; import header 0
0000012: 07                              ; string length
0000013: 696d 706f 7274 73    imports    ; import
  module name
000001a: 05                              ; string length
```

```
000001b:  6a73  4164  64            jsAdd   ; import field name
0000020:  00                                ; import kind
0000021:  00                                ; import signature
   index
```

Now, you can call the jsAdd function like you would other functions inside a WebAssembly module using the $i identifier.

We have explored how both the import and export sections are defined inside a WebAssembly module and how they help to import and export a function. Now, let's explore how to import and export values in and out of a WebAssembly module.

Globals

The globals section is where we can import and export values in and out of WebAssembly modules. In a WebAssembly module, you can import either mutable or immutable values from JavaScript. Additionally, WebAssembly also supports wasmValue, an internal immutable value inside the WebAssembly module itself.

Let's create a file called globals.wat and add the following contents to it:

```
$ touch globals.wat
(module
    (global $mutableValue (import "js" "mutableGlobal")
      (mut i32))
    (global $immutableValue (import "js"
      "immutableGlobal") i32)
    (global $wasmValue i32 (i32.const 10))
    (func (export "getWasmValue") (result i32)
      (global.get $wasmValue))
    (func (export "getMutableValue") (result i32)
      (global.get $mutableValue))
    (func (export "getImmutableValue") (result i32)
      (global.get $immutableValue))
    (func (export "setMutableValue") (param $v i32)
      (global.set $mutableValue
          (local.get $v)))
)
```

We created a module (`module`) and three global variables:

- `$mutableValue` – This value is imported from the `js` JavaScript module and the `mutableGlobal` variable. We also define the global variable to be of the `mut i32` type.

- `$immutableValue` – This value is imported from the `js` JavaScript module and the `immutableGlobal` variable. We also define the global variable to be of the `i32` type.

- `$wasmValue` – This is a global constant. We define the `global` keyword followed by the name of the global variable, `$wasmValue`, then the type of `i32`, and finally the actual value (`i32.const 10`).

> **Note**
> `$wasmValue` is immutable and cannot be exported to the external world.

Then, we have a set of functions that helps to get and set the global variables. `getWasmValue`, `getImmutableValue`, and `getMutableValue` get the values of the `wasmValue` global constant, the `immutableValue` global constant, and the `mutableValue` global variable, respectively.

Finally, a function that sets `mutableValue` to a new value is `setMutableValue`. `setMutableValue` takes in `param $v`, which sets the value to `$mutableValue`.

Let's use WABT to convert the WebAssembly text format into a WebAssembly module with the following command:

```
$ /path/to/wabt/bin/wat2wasm globals.wat
```

Create a `globals.html` with the following content:

```
// globals.html
<html>
    <head> </head>
    <body>
        <script>
            async function run() {   }
            run()
        </script>
    </body>
</html>
```

Let's define the run function inside `<script>`.

A `WebAssembly.Global` object represents a global variable instance, accessible from JavaScript and importable/exportable across one or more `WebAssembly.Module` instances. The `WebAssembly.Global` constructor expects a descriptor and value. The descriptor defines the type and mutability of the global variable defined:

> **Note**
> This global variable constructor provides an option to dynamically link multiple WebAssembly modules.

```
let immutableGlobal = new WebAssembly.Global({value:'i32',
  mutable:false}, 1000)
let mutableGlobal = new WebAssembly.Global({value:'i32',
  mutable:true}, 0)
```

We create two global values using the `WebAssembly.Global` constructor. They are `immutableGlobal` and `mutableGlobal`. The former is `mutable:false`, while the latter is `mutable:true`. So, we can change the value of the latter using `mutableGlobal.value` but not the former. If we try changing the value of `immutableGlobal`, then we will receive an error:

```
mutableGlobal.value = 1337  // valid.
immutableGlobal.value = 7331 // Error
```

After that, we fetch the `globals.wasm` WebAssembly module. Then, we instantiate `arrayBuffer` with the response and `arrayBuffer` with the `WebAssembly.instantiate` constructor. In addition to this, the `WebAssembly.instantiate` constructor accepts `importsObject`. We can send the JavaScript module via `importsObject`:

```
const response = await fetch('./globals.wasm')
const bytes = await response.arrayBuffer()
const wasm = await WebAssembly.instantiate(bytes, { js: {
  mutableGlobal, immutableGlobal } })
```

In this case, we are sending in the js module along with the mutableGlobal and immutableGlobal values. The wasm variable now holds the WebAssembly module. We invoke wasm.instance.exports to get all the exported functions from the WebAssembly module:

```
const {
    getWasmValue,
    getMutableValue,
    setMutableValue,
    getImmutableValue
} = wasm.instance.exports
```

getWasmValue, getMutableValue, setMutableValue, and getImmutableValue are the functions exported from the WebAssembly module.

The getWasmValue function returns the value of the wasmValue inside the WebAssembly module:

```
console.log(getWasmValue()) // 10
```

The getMutableValue and setMutableValue functions return and set the mutableGlobal field defined in JavaScript and passed into the WebAssembly module:

```
console.log(getMutableValue()) // 1337
setMutableValue(1338)
console.log(getMutableValue()) // 1338
```

Finally, we get the immutable value using the getImmutableValue function:

```
console.log(getImmutableValue()) // 1000
```

Let's run an example in the browser using the following command:

```
$ python -m http.server
```

Now, launch the URL http://localhost:8000/globals.html and open the developer tools.

The WebAssembly binary contains an import section. The import section has a unique identifier, 02, followed by the size of the section, which is 2b (which is 43 in decimal). The next 43 bits represent the import section.

`02` at the `000015` index represents the number of imports. Then, we have two sections that define the imported global functions:

```
; section "Import" (2)
0000013: 02                                      ; section code
0000014: 2b                                      ; section size
0000015: 02                                      ; num imports
```

Each global segment consists of the module string length and the module name, followed by the function string length and the function name. Finally, it has the kind of import, the type, and the mutability of the variable:

```
; import header 0
0000016: 02                                      ; string length
0000017: 6a73                          js        ; import module name
0000019: 0d                                      ; string length
000001a: 6d75 7461 626c 6547 6c6f 6261 6c
          mutableGlobal   ; import field name
0000027: 03                                      ; import kind
0000028: 7f                                      ; i32
0000029: 01                                      ; global mutability
   ; import header 1
000002a: 02                                      ; string length
000002b: 6a73                          js        ; import module name
000002d: 0f                                      ; string length
000002e: 696d 6d75 7461 626c 6547 6c6f 6261 6c
        immutableGlobal   ; import field name
000003d: 03                                      ; import kind
000003e: 7f                                      ; i32
000003f: 00                                      ; global mutability
```

After that, we have the `Global` section. The `Global` section has the unique section ID of 6. The next bit defines the size of the `Global` section, which is `06`.

After that, we have the number of globals available. The number of globals is `01`. This is because the other two globals are imported. The type, mutability, and value are the next 4 bytes:

```
; section "Global"  (6)
0000047: 06                              ; section code
0000048: 06                              ; section size
0000049: 01                              ; num globals
000004a: 7f                              ; i32
000004b: 00                              ; global mutability
000004c: 41                              ; i32.const
000004d: 0a                              ; i32 literal
000004e: 0b                              ; end
```

The first `function` body inside the code section looks as follows:

```
; function body 0
000009c: 04                              ; func body size
000009d: 00                              ; local decl count
000009e: 23                              ; global.get
000009f: 02                              ; global index
00000a0: 0b                              ; end
```

`function` is four bits in length. The first `00` says that the function has no local declaration. The next `23` is the opcode for getting the global value. The next `02` defines the index of the global value. Even though the preceding global section specifies there is only one global value, the entire module takes the imported globals into account. Since there are two imported global values, we index the local global values after the imported global value. So, the `$wasmValue` global has an index of 3. Finally, we end the function code with the `0b` opcode. Similarly, the second and third function bodies define how we get the other two imported global values:

```
; function body 3
00000ab: 06                              ; func body size
00000ac: 00                              ; local decl count
00000ad: 20                              ; local.get
00000ae: 00                              ; local index
```

```
00000af:  24                                      ; global.set
00000b0:  00                                      ; global index
00000b1:  0b                                      ; end
```

In function body 4, we set the global value using `global.set`, which has an opcode of 24.

We have explored how to import and export values in and out of WebAssembly modules. Now, let's explore the special `start` function in the WebAssembly module.

Start

Start is a special function that runs after the WebAssembly module is initialized. Let's take the same example that we used for the globals. We add the following content to `globals.wat`:

```
(module
    ; Code is elided
    (func $initMutableValue
          (global.set $mutableValue
              (i32.const 200)))
    (start $initMutableValue)
)
```

We define the `initMutableValue` function, which sets `mutableValue` to 200. After that, we add a start block, which starts with `startkeyword` followed by the name of the function.

> **Note**
> The function referenced at the start should not return any value.

Let's use WABT to convert the WebAssembly text format into a WebAssembly module with the following command:

```
$ /path/to/wabt/bin/wat2wasm globals.wat
```

Let's run the example in a browser using the following command:

```
$ python -m http.server
```

Now, launch the URL `http://localhost:8000/globals.html` and open the developer tools.

The start function is similar to other functions, except that it is not classified into any type. The types may or may not be initialized at the time of the function. The start section of a WebAssembly module points to a function index (the index of the location of the function section inside the function component).

The section ID of the start function is 8. When decoded, the start function represents the start component of the module:

```
; section "Start" (8)
0000085: 08                                ; section code
0000086: 01                                ; section size
0000087: 03                                ; start func index
```

> **Note**
>
> At this moment, tools such as webpack do not support the `start` function. The start section is rewritten into a normal function and then the function is invoked when the JavaScript is initialized by the bundler itself.

`start` is an interesting and useful function that enables setting up some values when the module is initialized to prevent unnecessary side effects that the module might cause. Now, let's explore the memory section. The memory section is responsible for transferring memory between JavaScript and WebAssembly.

Memory

Transferring data between JavaScript and WebAssembly is an expensive operation. In order to reduce the transfer of data between JavaScript and WebAssembly modules, WebAssembly uses `sharedArrayBuffer`. With `sharedArrayBuffer` both the JavaScript and WebAssembly modules can access the same memory and use it to share the data from one to the other.

The memory section of a WebAssembly module is a vector of linear memories. The linear memory model is a memory addressing technique in which the memory is organized in a single contiguous address space. It is also known as a flat memory model. The linear memory model makes it easier to understand, program, and represent the memory. But the linear memory model comes with a huge disadvantage of high execution time for rearranging elements in the memory and the wastage of memory space. Here, the memory represents a vector of raw bytes of uninterpreted data. They use resizable array buffers to hold the raw bytes of memory. We use `sharedArrayBuffers` for defining and maintaining this memory.

> **Note**
> It is important to note that this memory is accessible and mutable by JavaScript and WebAssembly.

We allocate the memory using the `WebAssembly.Memory()` constructor. The constructor can accept an argument that defines the initial and maximum value of memory, like so:

```
$ touch memory.html
$ vi memory.html
let memory = new WebAssembly.Memory({initial: 10, maximum: 100})
```

Here, we define that `WebAssembly.Memory` has an initial memory of `10` and a maximum memory of `100`. Then, we instantiate the WebAssembly module with the following code:

```
const response = await fetch('./memory.wasm')
const bytes = await response.arrayBuffer()
const wasm = await WebAssembly.instantiate(bytes, { js: { memory } })
```

Similar to the global example, here we are passing `importObject`, which takes in the `js` module and the memory object.

Let's create a new file called `memory.wat` and add the following content to it:

```
(module
    (memory (import "js" "memory") 1)
    (func (export "sum") (param $ptr i32) (param $len i32)
      (result i32)
```

```
(local $end i32)
(local $sum i32)
(local.set $end (i32.add (local.get $ptr)
  (i32.mul (local.get $len) (i32.const 4))))
(block $break (loop $top
    (br_if $break (i32.eq (local.get $ptr)
    (local.get $end)))
    (local.set $sum (i32.add (local.get $sum)
      (i32.load (local.get $ptr))))
    (local.set $ptr (i32.add (local.get $ptr)
      (i32.const 4)))
    (br $top)
))
(local.get $sum)
)
)
```

Inside the module, we import the memory from the js module with the name memory. After that, we define a function sum and export the function outside the module. The function accepts two parameters as arguments and returns an i32 as an output. The first parameter is named $ptr. It is a pointer to the index of where the value is present in sharedArrayBuffer. The next argument is $len, which defines the length of the shared memory.

Then, we create two local variables, $end and $sum. First, we set $end to the value of $ptr plus four times the value of $len. Then, we create a block and start a loop. The loop ends when the value of $end is equal to the value of $ptr. We then set the value of $sum by adding the existing value of $sum with the value of $ptr. Then, we increment $ptr to the next value. Finally, we exit the loop and return $sum.

The previous code is analogous to the following in JavaScript:

```
function sum(ptr, len) {
    let end = ptr + (len * 4)
    let tmp = 0
    while (ptr < end) {
        tmp = memory[ptr]
        ptr = ptr + 4
    }
    return tmp;
}
```

Let's go back to `memory.html` and initialize the buffer:

```
let i32Arr = new Uint32Array(memory.buffer)
for (var i = 0; i < 50; i++) {
    i32Arr[i] = i * i * i
}
```

We create an unsigned array using `Uint32Array` using the memory object we created. Then, we populate the array buffer with the cube of numbers from 1 to 50:

```
var sum = wasm.instance.exports.sum(0, 50)
console.log(sum) // 1500625
```

Finally, we call the sum inside the WebAssembly module and ask it to provide the sum of all the cubic numbers in the shared array buffer starting at 0 up to the length of 50.

Let's use WABT to convert the WebAssembly text format into a WebAssembly module with the following command:

```
$ /path/to/wabt/bin/wat2wasm memory.wat
```

Let's run an example in the browser using the following command:

```
$ python -m http.server
```

Now, launch the URL `http://localhost:8000/memory.html` and open the developer tools.

Memory sections are very useful when we have to transfer a large amount of data between two worlds. The memory sections make it easier to define, share, and access memory between the WebAssembly and JavaScript world.

Summary

In this chapter, we learned about the import, export, start, and memory sections in a WebAssembly module. We saw how they are structured and defined inside a WebAssembly module. Each of these sections carries one specific function and it is essential to understand, analyze, and debug the WebAssembly modules. In the next chapter, we will explore Binaryen.

6
Installing and Using Binaryen

During the compilation process, compiled languages produce their own **Intermediate Representation (IR)**. The compilers then optimize the IR to generate optimized code. Before passing it to LLVM, compilers should convert this IR into something that LLVM understands (LLVM IR). LLVM optimizes LLVM IR and produces native code (like the WebAssembly binary). These multiple IR generations and optimizations at different levels make the compilation process slower and not very effective. Binaryen tries to eliminate these multiple IR generations and uses its own IR.

(WebAssembly) Binary + Emscripten = Binaryen

> *Binaryen is a compiler and toolchain infrastructure library for WebAssembly, written in C++. It aims to make compiling to WebAssembly easy, fast, and effective.*
>
> *- Binaryen's GitHub repository (*`https://github.com/`
> `WebAssembly/binaryen`*)*

Binaryen uses its own version of IR. Binaryen's IR is a subset of WebAssembly. Thus, it makes compiling Binaryen to WebAssembly faster and easier. Binaryen's IR uses a compact data structure and is designed with modern CPU architecture in mind. That is, the WebAssembly binary can be generated and optimized in parallel using all the available CPU cores.

In addition to that, Binaryen's optimizer has many passes that can improve the code significantly. Binaryen's optimizer uses techniques such as local coloring to coalesce local variables, dead code elimination, and precomputing expressions wherever possible. Binaryen also provides a way to shrink the WebAssembly binary.

Binaryen is easy to use. It accepts the WebAssembly binary or even the control graph to generate a highly optimized WebAssembly binary. Binaryen also provides Binaryen.js, which enables the use of Binaryen from JavaScript. Similar to WABT, Binaryen includes a different set of tools that are useful while dealing with WebAssembly.

These toolchain utilities help in parsing the WebAssembly binary and then optimizing it further, and finally, emit a highly optimized WebAssembly binary (in other words, wasm-to-wasm optimizer), providing a polyfill for WebAssembly when the browser does not have WebAssembly support.

In this chapter, we will understand how to install and use various tools provided by Binaryen. Understanding Binaryen and the tools provided by it will help you to optimize the WebAssembly binaries in terms of performance and size. We will cover the following sections in this chapter:

- Installing and using Binaryen
- `wasm-as`
- `wasm-dis`
- `wasm-opt`
- `wasm2js`

Technical requirements

We'll be requiring Binaryen and Visual C++ installed. You can find the code files present in this chapter on GitHub at `https://github.com/PacktPublishing/Practical-WebAssembly`.

Installing and using Binaryen

In order to install Binaryen, first clone the repository from GitHub:

```
$ git clone https://github.com/WebAssembly/binaryen
```

After the repository is cloned, go into the folder:

```
$ cd binaryen
```

Linux/macOS

Generate the project build system by running the `cmake` command with the path to the `binaryen` folder:

```
$ cmake .
```

Next, build the project using the `make` command:

```
$ make .
```

This generates all the binaries in the `bin` folder.

Windows

For Windows, once the repository is cloned, we will create a `build` directory and go inside it:

```
$ mkdir build
$ cd build
```

By default, Windows does not have the cmake command available. Install the Visual C++ tools to make the `cmake` command available in the system. To install the Visual C++ tools, check out the following link: `https://docs.microsoft.com/ en-us/cpp/build/cmake-projects-in-visual-studio?view=msvc- 160&viewFallbackFrom=vs-2019`. Then, run the following command inside the `build` folder:

```
$ "<path-to-visual-studio-root>\Common7\IDE\CommonExtensions\
   Microsoft\CMake\CMake\bin\cmake.exe" ..
```

The preceding command will generate all the necessary build files in the `build` directory. Then, we can build the project using the `binaryen.vcxproj` file generated by `cmake`:

```
$ msbuild binaryen.vcxproj
```

The generated binary includes the following:

```
$ tree -L 1
├── wasm-as
├── wasm-ctor-eval
├── wasm-dis
├── wasm-emscripten-finalize
```

```
├── wasm-metadce
├── wasm-opt
├── wasm-reduce
├── wasm-shell
└── wasm2js
```

The various tools generated by Binaryen are as follows:

- `wasm-as` – This tool is similar to `wat2wasm` in WABT. This tool converts WebAssembly text format (`.wast`) into WebAssembly binary format (`.wasm`).

- `wasm-ctor-eval` – This tool executes C++ global constructors ahead of time and has them ready. This optimization speeds up the WebAssembly execution.

- `wasm-dis` – This tool is similar to `wasm2wat` in wabt. That is, it converts WebAssembly binary format (`.wasm`) into WebAssembly text format (`.wat`).

- `wasm-emscripten-finalize` – This tool performs Emscripten-specific transforms on the given `.wasm` files.

- `wasm-metadce` – This tool removes dead code in the provided WebAssembly binary.

- `wasm-opt` – This tool optimizes the provided WebAssembly binary.

- `wasm-reduce` – This tool reduces the given WebAssembly binary into a smaller binary.

- `wasm-shell` – This tool creates a shell that can load and interpret WebAssembly code.

- `wasm2js` – This tool is useful in polyfill. It converts WebAssembly into a JavaScript compiler.

- `binaryen.js` – A standalone JavaScript library that exposes Binaryen methods for creating and optimizing the WebAssembly modules. This JavaScript file is just like any other JavaScript file that can be loaded into the browser.

Now that we have built and generated the tools provided by Binaryen, let's explore the tools generated.

wasm-as

The `wasm-as` tool converts WAST into WASM. Let's look at the steps:

1. Let's create a new folder, called `binaryen-playground`, and go into the folder:

    ```
    $ mkdir binaryen-playground
    $ cd binaryen-playground
    ```

2. Create a `.wat` file called `add.wat`:

    ```
    $ touch add.wat
    ```

3. Add the following contents to `add.wat`:

    ```
    (module
        (func $add (param $x i32) (param $y i32)
            (result i32)
                (i32.add
                    (local.get $x)
                    (local.get $y)
                )
            )
        )
    ```

4. Convert the Web Assembly text format into a WebAssembly module, using the `wasm-as` binary:

    ```
    $ /path/to/build/directory/of/binaryen/wasm-as add.wat
    ```

    This generates the `add.wasm` file:

    ```
    00 61 73 6d 01 00 00 00 01 07 01 60 02 7f 7f 01
    7f 03 02 01 00 0a 09 01 07 00 20 00 20 01 6a 0b
    ```

    > **Note**
    > The size of the binary generated is just 32 bytes.

`wasm-as` first validates the given file (`.wat`) and then converts it into a `.wasm` file. To check various options supported by `wasm-as`, we can run the following command:

```
$ /path/to/build/directory/of/binaryen/wasm-as --help
wasm-as INFILE
Assemble a .wat (WebAssembly text format) into a .wasm
(WebAssembly binary
format)
Options:
  --version                    Output version information
     and exit
  --help,-h                    Show this help message and
     exit
  --debug,-d                   Print debug information to
     stderr
....
  --output,-o                  Output file (stdout if not
     specified)
  --validate,-v                Control validation of the
     output module
  --debuginfo,-g               Emit names section and debug
     info
  --source-map,-sm             Emit source map to the
     specified file
  --source-map-url,-su         Use specified string as
     source map URL
  --symbolmap,-s               Emit a symbol map (indexes
     => names)
```

If we need to generate the WebAssembly module file in a different name, we will use the `-o` option with the filename. For example, `wasm-as add.wat -o customAdd.wasm` will generate `customAdd.wasm`.

`wasm-as` also provides verbose output that clearly explains how the WebAssembly module is structured. In order to see the structure of the WebAssembly module, we run it with the `-d` option:

```
$ /path/to/build/directory/of/binaryen/wasm-as add.wat -d
Loading 'add.wat'...
```

```
s-parsing...
w-parsing...
Validating...
writing...
writing binary to add.wasm
Opening 'add.wasm'
== writeHeader
...
== writeTypes
...
== writeFunctionSignatures
...
== writeFunctions
...
finishUp
Done.
```

The previous output is a detailed description of how the binary is generated. First, it loads the given .wat file. After that, it parses and validates the file. Finally, it creates add.wasm and writes the header, the type, function signatures, and functions. While generating the binary, we can enable the compiler to include the new and shiny features and disable various existing features using the appropriate enable-* and disable-* options. Additionally, you can generate sourcemap using the --sm option.

Now that we have seen how to convert WAST to WASM, let's see how to convert WASM to WAST.

wasm-dis

The wasm-dis tool converts WAST into WASM. We will use the add.wasm file that we created in the previous example here. Let's look at the steps:

1. In order to convert the WebAssembly module into WebAssembly text format, using the wasm-dis binary, run the following command:

    ```
    $ /path/to/build/directory/of/binaryen/wasm-dis add.wasm
    -o gen-add.wast
    ```

2. We generate the `gen-add.wast` file using the `-o` option with the filename (`gen-add.wast`):

```
(module
(type $i32_i32_=>_i32 (func (param i32 i32)
  (result i32)))
(func $0 (param $0 i32) (param $1 i32) (result i32)
         (i32.add  (local.get $0)  (local.get $1) )
)
)
```

3. `wasm-dis` first validates the given file (`.wasm`) and then converts it into a `.wat` file. To check various options supported by `wasm-dis`, run the following command:

```
$ /path/to/build/directory/of/binaryen/wasm-dis --help
wasm-dis INFILE
Un-assemble a .wasm (WebAssembly binary format) into a
.wat (WebAssembly text
format)
Options:
  --version          Output version information and exit
  --help,-h          Show this help message and exit
  --debug,-d         Print debug information to stderr
  --output,-o        Output file (stdout if not specified)
  --source-map,-sm Consume source map from the specified
file to add location
```

4. `wasm-dis` also provides verbose output that clearly explains how the WebAssembly module is structured. In order to see the structure of the WebAssembly module, we run it with the `-d` option:

```
$ /path/to/build/directory/of/binaryen/wasm-dis
   add.wat -o gen-add.wast -d
parsing binary...
reading binary from add.wasm
Loading 'add.wasm'...
== readHeader
...
== readSignatures
```

```
...
== readFunctionSignatures
...
== processExpressions
...
== processExpressions finished
end function bodies
Printing...
Opening 'gen-add.wast'
Done.
```

The previous output is a detailed description of how the `.wast` file is generated. First, it loads the given `.wasm` file. After that, it parses and validates the file. Finally, it creates `gen-add.wast` after reading the header, type, function signatures, and functions.

While generating the file, we can enable the compiler to include the new and shiny features and disable various existing features using the appropriate `enable-*` and `disable-*` options, respectively.

Additionally, we can also input `sourcemap` using the `--sm <filename>` option.

Now that we have seen how to convert WASM to WAST, let's see how to optimize the WebAssembly binaries further.

wasm-opt

The `wasm-opt` tool is a `wasm-to-wasm` optimizer. It will receive a WebAssembly module as input and run transformation passes on it to optimize and generate the optimized WebAssembly module. Let's look at the steps:

1. Let's first create the `inline-optimizer.wast` file and add the following content:

```
(module
    (func $parent (export "parent") (result i32)
        (i32.add
            (call $child)
            (i32.const 13)
        )
    )
```

```
      )
      (func $child (result i32) (call $grandChild))
      (func $grandChild (result i32) (call
        $greatGrandChild))
      (func $greatGrandChild (result i32) (i32.const 7))
  )
```

2. To generate the WebAssembly module, we will run the following:

```
$ /path/to/bin/folder/of/binaryen/wasm-opt inline-
  optimizer.wast -o inline.wasm --print
(module
(type $0 (func (result i32)))
(export "parent" (func $parent))
(func $parent (; 0 ;) (type $0) (result i32)
(i32.add
(call $child)
(i32.const 13)
  )
)
(func $child (; 1 ;) (type $0) (result i32)
  (call $grandChild)
)
(func $grandChild (; 2 ;) (type $0) (result i32)
  (call $greatGrandChild)
)
(func $greatGrandChild (; 3 ;) (type $0) (result i32)
  (i32.const 7)
)
)
```

This will generate inline.wasm. The --print option prints the WebAssembly text format before converting it to the WebAssembly binary. We also passed in the -o option to output the WebAssembly module as inline.wasm:

```
60B inline-optimize.wasm
273B inline-optimize.wat
```

This generated binary of 60 bytes in memory.

3. We can further optimize the binary with the `--inlining-optimizing` option:

```
$ /path/to/bin/folder/of/binaryen/wasm-opt inline-
    optimizer.wast -o inline.wasm --print --inlining-
    optimizing
```

This will optimize the functions and inline the functions where the binary is called. Let's check what the file size generated is:

```
39B inline-optimize.wasm
273B inline-optimize.wat
```

The generated file is just 39 bytes, which is 35% less than the original binary.

4. To check various options supported by `wasm-opt`, run the following command:

```
/path/to/bin/folder/of/binaryen/wasm-opt -help
```

The `wasm-opt` tool helps us to optimize the WebAssembly binaries further. Let's next explore the `wasm2js` tool.

wasm2js

The `wasm2js` tool converts WASM/WAST files into JavaScript files. Let's look at the steps:

1. Create a file called `add-with-export.wast`:

```
$ touch add-with-export.wast
```

Then, add the following code:

```
(module
    (export "add" (func $add))
    (func $add (param $x i32) (param $y i32)
      (result i32)
        (i32.add
            (local.get $x)
            (local.get $y)
        )
    )
)
```

2. In order to convert the WebAssembly text format into JavaScript using wasm2js, run the following command:

```
$ /path/to/build/directory/of/binaryen/wasm2js add-
    with-export.wast
```

This will print out the generated JavaScript:

```
function asmFunc(global, env) {
var Math_imul = global.Math.imul;
var Math_fround = global.Math.fround;
var Math_abs = global.Math.abs;
var Math_clz32 = global.Math.clz32;
var Math_min = global.Math.min;
var Math_max = global.Math.max;
var Math_floor = global.Math.floor;
var Math_ceil = global.Math.ceil;
var Math_sqrt = global.Math.sqrt;
var abort = env.abort;
var nan = global.NaN;
var infinity = global.Infinity;
function add(x, y) {
  x = x | 0;
  y = y | 0;
  return x + y | 0 | 0;
}
return {
  "add": add
};
}
var retasmFunc = asmFunc({
    Math,
    Int8Array,
    Uint8Array,
    Int16Array,
    Uint16Array,
    Int32Array,
    Uint32Array,
```

```
        Float32Array,
        Float64Array,
        NaN,
        Infinity
    }, {
        abort: function() { throw new Error('abort'); }
    });
  export var add = retasmFunc.add;
```

The asmFunc function is defined. In asmFunc, we import the math functions from the global object. After that, we have an add function. The add function initializes x and y. The function returns the sum of two values. Finally, we return the add function.

> **Note**
>
> The generated JavaScript is asmjs and not normal JavaScript. We also imported a lot of functions from the global namespace in JavaScript into asmFunc.

The wasm2js tool makes it easy to generate JavaScript from a WebAssembly module. The generated JavaScript module is faster than its normal JavaScript counterpart. This can be used as a polyfill for browsers that do not support WebAssembly yet.

Summary

In this chapter, we have seen how to install Binaryen and what the various tools provided by the Binaryen toolkit are. Binaryen makes it easier to convert WebAssembly modules into various formats. It is an important tool that makes your WebAssembly journey easier and more efficient.

In the next chapter, we will start our Rust and WebAssembly journey.

Section 3: Rust and WebAssembly

This section introduces Rust and how easy it is to generate WebAssembly modules from Rust. You will learn how the Rust ecosystem made WebAssembly a first-class citizen and what the various tools available are and how to use them.

You will completely understand how to use Rust and WebAssembly together in a JavaScript application, as well as a little of WASI, by the end of this chapter.

This section comprises the following chapters:

- *Chapter 7, Integrating Rust with WebAssembly*
- *Chapter 8, Bundling WebAssembly Using wasm-pack*
- *Chapter 9, Crossing the Boundary between Rust and WebAssembly*
- *Chapter 10, Optimizing Rust and WebAssembly*

7
Integrating Rust with WebAssembly

Rust is a system-level programming language. Being a system-level programming language, Rust provides low-level memory management and the ability to represent data efficiently. Thus, it provides complete control to programmers and better performance.

In addition to this, Rust also provides the following:

- **A friendly compiler** – The Rust compiler is your companion when writing Rust. The compiler corrects you, guides you, and ensures that you write memory-safe code almost always.

- **The ownership model** – The ownership model ensures that we do not need garbage collection. This guarantees thread and memory safety in Rust.

- **Safety, speed, and concurrency** – Rust ensures safety and concurrency and makes you stay away from risks, crashes, and vulnerabilities.

- **A modern language** – Rust provides modern language syntax and the language is built to provide a better developer experience.

These features (along with thousands of others) ensure Rust is a general-purpose programming language. The highlight of the Rust language is that its compiler and community are always helpful.

Rust provides first-class support for WebAssembly. Rust's rich toolchain makes it easier to get started with WebAssembly. Rust does not need a runtime, which makes it a perfect candidate for WebAssembly. In this chapter, we will see how to install Rust and explore various ways to convert Rust into a WebAssembly module. We will cover the following sections in this chapter:

- Installing Rust
- Converting Rust into WebAssembly via `rustc`
- Converting Rust into WebAssembly via Cargo
- Installing wasm-bindgen
- Converting Rust into WebAssembly via `wasm-bindgen`

Now let's hack into the Rust and WebAssembly world.

Technical requirements

You can find the code files present in this chapter on GitHub at `https://github.com/PacktPublishing/Practical-WebAssembly`.

Installing Rust

Rust is a compiled language and its compiler is called the **Rust compiler** (**rustc**). Rust also has its own package manager, called **Cargo**. Cargo is similar to npm for Node.js. Cargo downloads package dependencies and builds, compiles, packs, and uploads the artifacts into crates (Rust's version of packages).

The Rust language provides an easy way to install and manage Rust via `rustup`. `rustup` helps to install, update, and remove `rustc`, Cargo, and `rustup` itself. It makes it easy to install and manage various versions of Rust.

Let's install Rust using the `rustup` tool and see how we can manage Rust versions using `rustup`.

In Linux or macOS, use the following command:

```
$ curl https://sh.rustup.rs --sSf | sh
```

The script will download and install the Rust language. Both `rustc` and Cargo are installed in `~/.cargo/bin` and delegate any access to the underlying toolchain.

For Windows, download and install the binaries from here: `https://forge.rust-lang.org/infra/other-installation-methods.html`. Both `rustc` and Cargo are installed in the `users` folder.

> **Note**
>
> You will require C++ build tools for Visual Studio 2013 or later. You can install them from `https://visualstudio.microsoft.com/downloads/`.

Once the installation is completed successfully, you can check it by running the following command:

```
$ rustc -version
rustc 1.58.1 (db9d1b20b 2022-01-20)
```

`rustup` is a toolchain multiplexer. It installs and manages many Rust toolchains and proxies them through the single set of tools installed at `.cargo/bin` in the home directory. Once `rustup` is installed, we can easily manage the `rustc` and `cargo` compilers. `rustup` also makes it easy to switch between nightly, stable, and beta versions of Rust.

Rust provides WebAssembly compilation support in its stable version. We will also switch to the nightly build to make sure we get all the latest benefits.

To switch to the nightly version, we have to run the following command:

```
$ rustup default nightly
```

This command will switch the default `rustc` to the nightly version. The `rustc` proxy in `~/.cargo/bin` will run the nightly compiler instead of the stable compiler.

To update to the latest version of nightly, we can run the following:

```
$ rustup update
```

Once successfully updated, we can check the current version installed by running the following:

```
$ rustc --version
rustc 1.55.0 (c8dfcfe04 2021-09-06)
```

Rust supports WebAssembly as a first-class citizen. Thus, `rustc` is capable of compiling Rust code into WebAssembly modules. Let's see how to convert Rust into WebAssembly via `rustc`.

Converting Rust into WebAssembly via rustc

Rust uses the LLVM compiler we'll create now to generate machine-native code. `rustc` uses LLVM's capability to convert the native code into a WebAssembly module. We installed Rust in the previous section; let's start converting Rust into a WebAssembly module using `rustc`.

We will start with Hello World:

1. Let's create a file called `hello_world.rs`:

   ```
   $ touch hello_world.rs
   ```

2. Spin up your favorite editor and start writing the Rust code:

   ```
   fn main() {
   println!("Hello World!");
   }
   ```

   We have defined a `main` function. Similar to C, `main` is a special function that marks the entry point to a program after it has been compiled as an executable.

   `fn` is the function keyword in Rust. `main()` is the function name.

   `println!` is the macro. Macros in Rust allow us to abstract code at a syntactic level. A macro invocation is shorthand for an "expanded" syntactic form. This expansion happens early on in compilation, before any static checking.

   > **Note**
   > Macros are an interesting feature but explaining them is beyond the scope of this book. You can find more information here: `https://doc.rust-lang.org/book/ch19-06-macros.html`.

3. We pass in the `Hello World!` string to the `println!` macro function. We can compile and generate the binary by running the following:

```
$ rustc hello.rs
```

4. This will generate a `hello` binary. We can execute the binary and that will print `Hello World!`:

```
$ ./hello
Hello World!
```

5. Now, compile Rust into a WebAssembly module with `rustc`:

```
$ rustc --target wasm32-unknown-emscripten
  hello_world.rs -o hello_world.html
```

This will generate the WebAssembly module.

> **Note**
> Install `wasm32-unknown-emscripten` target using `$ rustup target add wasm32-unknown-emscripten`

6. Let's run the generated code in the browser using the following command:

```
$ python -m http.server
```

Open your browser and head over to `http://localhost:8000/hello_world.html`. Open the developer console to see `Hello World!` printed in it.

To convert Rust into a WebAssembly module, we have used the `--target` flag. This flag instructs the compiler to compile and build the binary such that it runs on the provided runtime.

We passed in `wasm32-unknown-emscripten` as a value to the `--target` flag.

`wasm32` indicates that the address space is 32 bits large. `unknown` tells the compiler that you don't know the system that you are compiling to. `emscripten` at the end notifies the compiler that you are targeting.

So, with the `wasm32-unknown-emscripten` value, the compiler will compile on almost any machine but run only on the Emscripten runtime. Then, we specify the input file that needs to be compiled into the WebAssembly module. Finally, we specify the output with a `-o` flag.

It is important to understand what `rustc` does.

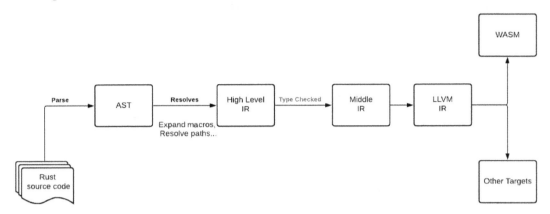

Figure 7.1 – Rust compilation steps

`rustc` first parses the input and produces the **Abstract Syntax Tree** (**AST**). Once the AST is generated, the compiler then recursively resolves the paths, expanding macros and other references. Once the AST is completely resolved, it will be converted into the **High-level Intermediate Representation** (**HIR**). This intermediate representation is like a desugared variant of AST.

The HIR is then analyzed for type checking. After type checking, the HIR is postprocessed and converted into the **Middle Intermediate Representation** (**MIR**). From MIR, the compiler generates the **LLVM Intermediate Representation** (**LLVM IR**). After that, LLVM does the required optimizations on them.

Now, with LLVM IR, it is easier to convert LLVM IR into WebAssembly modules. This is similar to how Emscripten converts C or C++ code into WebAssembly modules.

> **Note**
>
> Since we are using the `wasm32-unknown-emscripten` flag here, we need `emcc` to be available for converting the LLVM IR generated from Rust code into a WebAssembly module.

We have seen how to use `rustc` to generate WebAssembly modules. It uses Emscripten behind the scenes to create them. But Rust provides another abstraction to generate WebAssembly modules, via Cargo.

In the next section, we will see how to convert Rust into WebAssembly using Cargo.

Converting Rust into WebAssembly via Cargo

Cargo makes it easier to create, run, download, compile, test, and run your project. The cargo command provides a wrapper that calls the rustc compiler to start the compilation. In order to create WebAssembly modules using Rust's toolchain, we will be using a different target, wasm32-unknown-unknown.

The wasm32-unknown-unknown target adds zero runtime and toolchain footprint. wasm32 makes the compiler assume that only the wasm32 instruction set is present. The first unknown in unknown-unknown indicates the code can compile on any machine and the second indicates the code can run on any machine.

To see it in action, let's create a new project with Cargo:

```
$ cargo new --lib fib_wasm
  Created library `fib_wasm` package
```

A new project called fib_wasm is created. The new option creates a Rust project. The --lib flag informs Cargo to create a new library project rather than the default binary project.

The binary project will produce the executable binary. The library project will create the library module.

Spin up your favorite text editor and replace the contents of src/lib.rs with the following:

```
#[no_mangle]
  fn add(x: i32, y:i32) -> i32 {    x + y}
```

#[no_mangle] is a kind of annotation. This annotation informs the compiler not to mangle the names when generating the library.

Then, we define the add function. The add function takes in two parameters, x and y. We define their types with i32 following the variable and a colon (:). Finally, we define their return type using -> i32.

The function body has just x + y. Note, in Rust we do not need return keyword and ; at the end of the last statement, this shorts to return.

Cargo also generates Cargo.toml. This file holds all the meta-information about the project, how to compile the Rust code, and their dependencies.

The `Cargo.toml` file looks like this:

```
[package]
name = "fib_wasm"
version = "0.1.0"
authors = ["Sendil Kumar"]
edition = "2018"
```

It defines the package name, version, authors, and edition of Rust we are using.

Here, we have to instruct the compiler what type of crate we are compiling. We can specify it under the `[lib]` segment and with the `crate-type` property.

Open `Cargo.toml` and append the `crate-type` information inside:

```
[package]
name = "fib_wasm"
version = "0.1.0"
authors = ["Sendil Kumar"]
edition = "2018"
[lib]
crate-type = ["cdylib"]
```

`cdylib` here specifies a dynamic system library will be produced. This dynamic system library is used when the library has to be loaded from another language.

> **Note**
>
> Install wasm32-unknown-unknown target using `$ rustup target add wasm32-unknown-unknown`

Let's compile Rust into WebAssembly modules:

```
$ cargo build --target wasm32-unknown-unknown
```

This will invoke `rustc` with the specified target. That will generate the WebAssembly module inside `/target/wasm32-unknown-unknown/`. Now, in order to run the WebAssembly module on the browser, let's manually create the HTML file and load it using JavaScript.

Let's create an HTML file:

```
$ touch index.html
```

Add the following content to the file:

```
<script>
(async () => {
  const bytes = await fetch("target/wasm32-unknown-
    unknown/debug/fib_wasm.wasm");
  const response = await bytes.arrayBuffer();
  const result = await WebAssembly.instantiate(response, {});
    console.log(result.instance.exports.add(10,3));
})();
</script>
```

We defined the script inside the `<script>` tag. In HTML, we define JavaScript inside the `<script>` tag. We have added the **Immediately Invoked Function Execution (IIFE)** block with the `async` keyword. The `async` keyword specifies the function is asynchronous.

First, we fetch the WebAssembly module. The WebAssembly module is generated in the `target/wasm32-unknown-unknown/debug/` folder with the same name as the package name defined in `Cargo.toml`.

The `await` keyword ensures the execution is awaited until we fetch the entire WebAssembly module.

We then convert the collected bytes (from the fetch call) using `bytes.arrayBuffer()`. The `response` object will now have the WebAssembly module inside `ArrayBuffer`.

We then instantiate the bytes array using the `WebAssembly.instantiate` function. The `result` object contains the entire WebAssembly module.

The WebAssembly module `result` contains the `instance` property. The instance has the `exports` property. The `exports` property holds all the functions exported by the WebAssembly module.

Since we added `#[no_mangle]`, the exported function name is not changed. Hence, the `exports` property has the `add` function defined in it.

We have used async-await here to make the syntax more elegant and contextually easier to understand.

As expected, the preceding code will give an output of 13. You can check the output in the browser console.

Here, the `cargo build` command invokes `rustc` and compiles the Rust code into MIR and then into LLVM IR. The generated LLVM IR is then converted into a WebAssembly module. Let's make this function a bit more complicated. We can create a Fibonacci number generator with Rust and run the WebAssembly Module on the browser:

1. Open `src/lib.rs` and replace everything with the following content:

```
#[no_mangle]
fn fibonacci(num: i32) -> i32 {
    match num {
        0 => 0,
        1 => 1,
        _ => fibonacci(num-1) + fibonacci(num-2),
    }
}
```

Build it using `cargo build --target wasm32-unknown-unknown`.

2. Then, replace `index.html` such that we call the Fibonacci instead of the `add`:

```
<script>
  (async () => {
    const bytes = await fetch("target/wasm32-unknown-
    unknown/debug/fib_wasm.wasm");
    const response = await bytes.arrayBuffer();
    const result = await WebAssembly.
      instantiate(response,
      {});
    result.instance.exports.fibonacci(20);
})();
</script>
```

3. Now, spin up the HTML server and check the browser's console for the Fibonacci value.

So far, we have seen simple examples. But how can we pass functions and classes from JavaScript into WebAssembly and the other way around? In order to do more advanced bindings, Rust provides us with `wasm-bindgen`.

Installing wasm-bindgen

wasm-bindgen is used to bind entities from Rust into JavaScript and vice versa.

wasm-bindgen makes it more natural to import exported entities from Rust into JavaScript. JavaScript developers will find that wasm-bindgen's use of WebAssembly is similar to JavaScript.

This enables the use of richer and easier APIs while converting Rust into a WebAssembly module. wasm-bindgen uses these features and provides a simple API to use. It ensures that there is high-level interaction happening between wasm modules and JavaScript.

wasm-bindgen provides a channel between JavaScript and WebAssembly to communicate something other than numbers, such as objects, strings, and arrays.

To install wasm-bindgen-cli, use the following cargo command:

```
$ cargo install wasm-bindgen-cli
```

Once successfully installed, let's run the wasm-bindgen CLI:

```
$ wasm-bindgen  --help
Generating JS bindings for a wasm file
Usage:
    wasm-bindgen [options] <input>
    wasm-bindgen -h | --help
    wasm-bindgen -V | --version
Options:
    -h --help               Show this screen.
    --out-dir DIR           Output directory
    --out-name VAR          Set a custom output filename
        (Without extension. Defaults to crate name)
    --target TARGET         What type of output to generate,
        valid
        values are [web, bundler, nodejs, no-modules],
            and the default is [bundler]
    --no-modules-global VAR  Name of the global variable to
        initialize
    --browser               Hint that JS should only be
        compatible with a browser
    --typescript            Output a TypeScript definition
```

`file (on by default)`	
`--no-typescript`	`Don't emit a *.d.ts file`
`--debug`	`Include otherwise-extraneous`
`debug checks in output`	
`--no-demangle`	`Don't demangle Rust symbol names`
`--keep-debug`	`Keep debug sections in wasm files`
`--remove-name-section`	`Remove the debugging ` + "`name`"
`section of the file`	
`--remove-producers-section`	`Remove the telemetry`
"`producers`" + ` section`	
`--encode-into MODE`	`Whether or not to use`
`TextEncoder#encodeInto,`	
`valid values are [test,z always, never]`	
`--nodejs`	`Deprecated, use ` + "`--target nodejs`"
`--web`	`Deprecated, use ` + "`--target web`"
`--no-modules`	`Deprecated, use ` + "`--target`"
`no-modules`"	
`-V --version`	`Print the version number of`
`wasm-bindgen`	

Let's take a look at the various options `wasm-bindgen` supports.

To generate the file in a particular directory and with a particular name, the tool has `--out-dir` and `--out-name`, respectively. To reduce or optimize the generated WebAssembly module size, `wasm-bindgen` has the following flags:

- `--debug`: The `--debug` option includes extra debugging information in the generated WebAssembly module. This will increase the size of the WebAssembly module but it is useful in development.

- `--keep-debug`: This WebAssembly module may or may not have custom sections. These custom sections can be used to hold the debugging information. These custom sections will be helpful while debugging the application (such as in browser developer tools). This will increase the size of the WebAssembly module. This is useful in development.

- `--no-demangle`: This flag tells `wasm-bindgen` not to demangle the Rust symbol names. Demangling allows the end user to use the same name that they have defined in the Rust file.

- `--remove-name-section`: This will remove the debugging name section of the file. We will see more about the various sections in a WebAssembly module later. This will decrease the size of the WebAssembly module.

- `--remove-producers-section`: WebAssembly modules can have a producer section. This section will hold information about how the file was produced or who produced the file. By default, producer sections are added to a generated WebAssembly module. With this flag, we can remove it. It saves a few more bytes.

`wasm-bindgen` provides options to generate the binding file for both Node.js and the browser environment. Let's see those flags:

- `--nodejs`: This generates output that only works for Node.js. No ESModules.

- `--browser`: This generates output that only works for the browser. With ESModules.

- `--no-modules`: This generates output that only works for the browser. No ESModules. Suitable for browsers that don't support ESModules yet.

The type definition files (`*.d.ts`) can be switched off by using the `--no-typescript` flag.

Now we have installed `wasm-bindgen`, let's take it for a spin.

Converting Rust into WebAssembly via wasm-bindgen

Let's start with a Hello World example with `wasm-bindgen`:

1. Create a new project with Cargo:

```
$ cargo new --lib hello_world
  Created library `hello_world` package
```

This will create a new Rust project with all the necessary files.

2. Open the project in your favorite editor. Open the `Cargo.toml` file to add `crate-type` and add the `wasm-bindgen` dependency:

```
[package]
name = "hello_world"
version = "0.1.0"
authors = ["Sendil Kumar"]
edition = "2018"
[lib]
crate-type = ["cdylib"]
[dependencies]
wasm-bindgen = "0.2.38"
```

3. We define the dependency under the `[dependencies]` table in the `toml` file. Open the `src/lib.rs` file and replace the contents with the following:

```
use wasm_bindgen::prelude::*;
#[wasm_bindgen]
pub fn hello_world() -> String {
"Hello World".to_string()
}
```

We import the wasm_bindgen library using use `wasm_bingen::prelude::*` and then annotate the function with # `[wasm_bindgen]`. The `hello` function returns `String`.

To generate the WebAssembly module, we will first run the following command:

```
$ cargo build --target=wasm32-unknown-unknown
```

This will generate the WebAssembly module. But this module cannot run by itself. WebAssembly only supports passing numbers between the native code and JavaScript. But we are returning a `String` here.

In order to pass any value (other than numbers), we need to create a binding JavaScript file. This binding file is nothing more than a translator that translates the `String` (and other types) into `start`, `length`, `arrayBuffer`.

In order to generate the binding files, we need to run the `wasm-bindgen` CLI tool on the generated WebAssembly module:

```
$ wasm-bindgen target/wasm32-unknown-
  unknown/debug/hello_world.wasm --out-dir .
```

We run `wasm-bindgen` and pass it to the generated `target/wasm32-unknown-unknown/debug/hello_world.wasm` WebAssembly module. The `--out-dir` flag tells the `wasm-bindgen` CLI tool where to save the generated files. Here, we are asking for the files to be generated in the current folder.

We can see the files that are generated inside the folder:

```
$ ls -lrta
-rw-r--r-- 1 sendilkumar staff 1769 hello_world.js
-rw-r--r-- 1 sendilkumar staff 88 hello_world.d.ts
-rw-r--r-- 1 sendilkumar staff 227 hello_world_bg.d.ts
-rw-r--r-- 1 sendilkumar staff 67132 hello_world_bg.wasm
```

The `cargo build` command generates the WebAssembly module. The `wasm-bindgen` CLI takes this WebAssembly module as input and generates the necessary bindings. The size of the binding JavaScript file is around 1.8 KB.

The generated files are as follows:

- The WebAssembly module (`hello_world_bg.wasm`)
- The JavaScript binding file (`hello_world.js`)
- The type definition file for the WASM (`hello_world.d.ts`)
- The type definition file for the JavaScript (`hello_world_bg.d.ts`)

The JavaScript binding file is enough for us to load and run the WebAssembly module.

> **Note**
> There is also a TypeScript file generated.

Let's now check what the binding file contains:

```
import * as wasm from './hello_world_bg.wasm';
```

The binding file imports the WebAssembly module:

```
const lTextDecoder = typeof TextDecoder === 'undefined' ?
  require('util').TextDecoder : TextDecoder;
let cachedTextDecoder = new lTextDecoder('utf-8');
```

It then defines `TextDecoder`, to decode the string from the shared `ArrayBuffer`.

> **Note**
>
> Since there are no input arguments available, there is no need for `TextEncoder` (that is, to encode the string from JavaScript into shared memory). `wasm-bindgen` will generate only the necessary things inside the binding file.

Modern browsers have built-in `TextDecoder` and `TextEncoder` support. `wasm-bindgen` checks and uses the decoder if present; otherwise, it loads it using polyfill.

The shared memory between JavaScript and the WebAssembly module need not be initialized every time. We can initialize it once and use it throughout the lifetime of the execution. We have the following two methods to load the memory once and use it throughout the lifetime of the execution:

```
function getUint8Memory() { ... }
function getUint32Memory() { ... }
```

Then, we get a `String` from Rust to JavaScript. This `String` is passed via the shared memory. So, we can use the pointer to the offset and the length of the `String` to retrieve it. The following function is used for retrieving the `String` from the WebAssembly module:

```
function getStringFromWasm(ptr, len) { .... }
```

We define the heap at the very end. This is where we will store all the JavaScript variables referenceable from the WebAssembly module. The `__wbindgen_object_drop_ref` function is used to free up the slot occupied by the JavaScript reference count.

Finally, we have the `hello` function:

```
export function hello() {
    const retptr = globalArgumentPtr();
    wasm.hello_world(retptr);
    const mem = getUint32Memory();
    const rustptr = mem[retptr / 4];
    const rustlen = mem[retptr / 4 + 1];
    const realRet = getStringFromWasm(rustptr,
      rustlen).slice();
    wasm.__wbindgen_free(rustptr, rustlen * 1);
    return realRet;
}
```

The `hello` function is exported. We first get the pointer for the argument. This pointer refers to a location in the shared array buffer. Then, we call the `hello` function in the WebAssembly module.

Note that we are passing in a (pointer to the) argument here. But we have defined the function without any arguments on the Rust side. We will briefly see how `rustc` has rewritten the code.

Then, we get the shared memory. Note that this is a 32-bit array. We get the pointer in which the result is stored and the length of the output string. Note that these are stored successively.

Finally, we will get the string from `rustptr` and `rustlen`. Once we have received the output, we will clear the allocated memory using `wasm.__wbindgen_free`.

To understand what happens on the Rust side, let's use the `cargo-expand` command to expand the macro and see how the code is generated.

> **Note**
> Check `https://github.com/dtolnay/cargo-expand` for how to install `cargo-expand`. It is not mandatory for the course of this book. But cargo-expand will help you understand what `wasm-bindgen` actually generates.

Open your terminal, go to the project's base directory, and run the following command:

```
cargo expand --target=wasm32-unknown-unknown > expanded.rs
```

The preceding command will create a file called expanded.rs. If you take a look in the file generated, you will see how the simple #[wasm_bindgen] annotation changes the verbose part of exposing the function. The wasm-bindgen adds all the necessary metadata that is required for the compiler to convert Rust code into a WebAssembly module. To load and run the generated files, we can use bundlers such as webpack or Parcel. We will see how these bundlers help in more detail in later chapters. For now, let's see how to run and load the generated files:

> **Note**
>
> The following setup is common and we will refer to it as the "default" webpack setup in future examples.

Create a webpack-config.js file to tell webpack how to handle the files.

```
const path = require('path');
const HtmlWebpackPlugin = require('html-webpack-plugin');
module.exports = {
    entry: './index.js',
    output: {
        path: path.resolve(__dirname, 'dist'),
        filename: 'bundle.js',
    },
    plugins: [
        new HtmlWebpackPlugin(),
    ],
    mode: 'development'
};
```

This is a standard webpack configuration file with an HTMLWebpackPlugin plugin. This plugin helps us to generate a default index.html instead of manually creating it.

Let's add a package.json file to bundle the dependencies for running webpack:

```
{
    "scripts": {
        "build": "webpack",
        "serve": "webpack-dev-server"
    },
    "devDependencies": {
```

```
        "html-webpack-plugin": "^3.2.0",
        "webpack": "^4.29.4",
        "webpack-cli": "^3.1.1",
        "webpack-dev-server": "^3.1.0"
    }
}
```

Create an `index.js` file to load the binding JavaScript, which in turn loads the WebAssembly module generated:

```
import("./hello_world").then(module => {
    console.log(module.hello_world());
});
```

Now, head over to the terminal and install the npm dependencies using the following command:

```
$ npm install
```

Run `webpack-dev-server`:

```
$ npm run serve
```

Go to the URL `http://localhost:8080` and open the developer console in the browser to see "Hello World" printed.

Summary

In this chapter, we saw how to install Rust using `rustup`. `rustup` helps us to install, update, remove, and switch different versions of Rust. We saw how `rustc` works and then converted Rust into WebAssembly using `rustc`. After that, we explored Cargo, the package manager for Rust. Finally, we installed `wasm-bindgen` and compiled Rust code into a WebAssembly module using `wasm-bindgen`.

In the next chapter, we will explore what `wasm-pack` is and how it helps to build and pack WebAssembly modules.

8
Bundling WebAssembly Using wasm-pack

JavaScript is omnipresent, but being everywhere is both an advantage and a disadvantage. There are many different ecosystems that have various standards and purposes. Building a unique solution for all ecosystems is not practical.

Despite all this, the JavaScript community is doing a wonderful job here. The effort from the community makes JavaScript one of the go-to languages. For a language as versatile as JavaScript, there will be some weird corners (which of course every language has). When you are writing JavaScript, these need extra care and attention.

JavaScript is dynamically typed. This makes it difficult (almost impossible) to avoid runtime exceptions. While TypeScript, Flow, and Elm try to provide a (typed) superset on JavaScript's dynamic types, they cannot completely fix the underlying problem.

For any language to grow, it has to evolve fast, which JavaScript does. Evolving fast without breaking the existing usage is also important and JavaScript provides polyfills to make it backward compatible.

But creating polyfills is a mundane task. There are various other mundane steps, such as bundling and packaging libraries, minifying bundles, and lazy-loading libraries, to name a few. Bundlers provide a solution to most of them. They act as a compiler for the frontend.

So far, we have seen how Rust makes it easy to create and run WebAssembly modules. In this chapter, we will explore `wasm-pack`, a tool that makes it easier to pack and publish WebAssembly modules. We will cover the following sections in this chapter:

- Bundling WebAssembly modules with webpack

- Bundling WebAssembly modules with Parcel

- Introducing `wasm-pack`

- Packing and publishing using `wasm-pack`

Technical requirements

You can find the code files present in this chapter on GitHub at `https://github.com/PacktPublishing/Practical-WebAssembly`.

Bundling WebAssembly modules with webpack

webpack is a static module bundler for modern JavaScript applications. So, what does it do?

You can consider webpack as an informal compiler for the frontend. webpack takes in an entry point of an application, slowly runs through the modules, and builds a dependency graph. The dependency graph holds all the modules. These modules are necessary for the application to run.

Once the dependency graph is built, webpack outputs one or more bundles. Webpack is very flexible, helping us to bundle or package JavaScript as we need it and the options are provided in the webpack configuration. Based on the provided options, webpack creates the output.

Well, that sounds simple, right?

It was that simple a few years ago when the only library that we needed was jQuery.

But due to JavaScript's rapid evolution, there are a lot of different things happening now. The underlying runtime is not the same. There are three different browser engines and various targets.

Browser engines evolve at different speeds and browsers support various versions of JavaScript. In some workplace machines, upgrading browsers to the latest version is prohibited. This means the running JavaScript application needs tweaking and polyfills at various times.

The underlying target system needs a certain tweak to make your JavaScript code run. Doing all this by hand will take a long time to complete and will be error-prone.

There are various flavors of JavaScript, including TypeScript and CoffeeScript. They are different, but they will compile down to JavaScript before running. Browser-based development needs CSS, SCSS, SASS, and LESS. Supporting all those variations and compiling them manually after every change is not an easy deal.

JavaScript's answer to all this is bundlers. Whether you hate them or love them, bundlers reduce the overload and remove the clutter when developing with JavaScript.

webpack provides a solution to all these problems and more.

webpack is a tool built for bundling JavaScript applications. It comes with loaders and plugins that will help to convert, add, remove, and manipulate the output bundles. The most interesting part of webpack is its loaders and plugins, which propel the ability of webpack to the fullest.

Loaders allow us to load or import a Rust, CSS, or TypeScript file like any other module inside JavaScript. webpack then takes care of producing the bundle that will support the target environment as specified.

Plugins allow us to optimize and manage the bundles produced. It is important to note that webpack is built entirely on top of this plugin system.

How does webpack help with WebAssembly?

webpack internally depends on the `webassemblyjs` library. So, all the applications that use webpack are already WebAssembly-ready. All you have to do is start loading the WebAssembly file as normal JavaScript and webpack will take care of the rest.

In the webpack configuration, we will define the entry point. webpack then loads the entry file. The `import` statements in the entry file are loaded as a module based on JavaScript's module resolution algorithm. If the imported module is a WebAssembly module, it gets the module's content and hands it over to the `webassemblyjs` compiler.

The compiler is responsible for parsing and mutating the WebAssembly modules.

> **Did You Know?**
>
> `webassemblyjs` can parse the WebAssembly text format and WebAssembly binary format out of the box.

The compiler generates the **abstract syntax tree (AST)**. The generated AST is then validated. Once the validation is successful, any custom sections in the WebAssembly module are removed.

The custom section is a section inside the WebAssembly module where users can store custom information about the WebAssembly module. This information may include names of the function and local variables. Browsers may then use this information to have a better debugging process.

webpack also does not support the start section. The start section is a section in a WebAssembly module that will be called as soon as the WebAssembly module is loaded.

Instead, webpack creates a function and calls it after the WebAssembly module is loaded. `webassemblyjs` removes the start section and converts the start function into a normal function on the WebAssembly module. Then, webpack takes care of generating the wrapper that calls the function as soon as the module is loaded.

Finally, `webassemblyjs` is also responsible for optimizing the binary and eliminating dead code from the WebAssembly module.

`webassemblyjs` comes with an interpreter and CLI, which makes it easy to experiment with the WebAssembly modules.

Code, refresh, and repeat.

This was the workflow for web development for a long time. Live reloading provides an extra pair of hands to web developers. Live reloading can automatically compile and reload the changes once the code is saved. The code can be shared across multiple devices, factors, and orientations. Interactions in one place can automatically be synchronized with other devices. While the web provides a medium to deliver software easily, it comes in various forms. These forms are feature phones, smartphones, tablets, laptops, computers, ultra-wide monitors, 360-degree virtual worlds, and so on. Supporting all or some of them is an uphill task. Live reloading works like an extra pair of hands.

webpack provides multiple options to add live reloading to your application. It provides plugins for live reloading tools, such as BrowserSync. The webpack ecosystem also provides a watch mode in its configuration.

The watch mode, once enabled, looks for any changes that happen in the source file and its directory. Once the changes are detected, it will recompile automatically. But watch mode is for recompiling input into output.

The automatic reloading of web pages is provided by a library called webpack-dev-server. webpack-dev-server is an in-memory web server. The contents are generated and are placed in memory rather than in actual files in the filesystem.

In addition to that, webpack-dev-server also supports Hot Module Replacement. This allows the server to patch only the changes in the browser rather than doing a full page refresh.

Let's see how we can enable live reloading in a WebAssembly project:

1. First, we will create a new Rust project:

```
$ cargo new --lib live_reload
    Created library `live_reload` package
```

2. Once the project is created, open it in your favorite editor. To define the wasm-bindgen dependency for the project, open the Cargo.toml file:

```
[package]
name = "live_reload"
version = "0.1.0"
authors = ["Sendil Kumar"]
edition = "2018"

[lib]
crate-type = ["cdylib"]

[dependencies]
wasm-bindgen = "0.2.38"
```

First, add the [lib] section and add crate-type = ["cdylib"]. With the crate-type option, we are instructing the compiler that the library is dynamic. After that, add the wasm-bindgen dependency to the [dependencies] tag.

3. Then, open the src/lib.rs file and replace the contents with the following:

```
use wasm_bindgen::prelude::*;

#[wasm_bindgen]
```

```
pub fn hello_world() -> String {

"Hello World".to_string()
}
```

4. We will reuse the simple Hello World example from previous chapters here. Build the WASM module using the following:

```
$ cargo build --target wasm32-unknown-unknown
$ wasm-bindgen target/wasm32-unknown-
  unknown/debug/live_reload.wasm --out-dir .
```

5. Then create a webpack.config.js file to instruct webpack on how to handle and compile the files:

```
const path = require('path');
const HtmlWebpackPlugin = require('html-webpack-
  plugin');

module.exports = {
    entry: './index.js',
    output: {
        path: path.resolve(__dirname, 'dist'),
        filename: 'bundle.js',
    },
    plugins: [
        new HtmlWebpackPlugin(),
    ],
    experiments: {
        syncWebAssembly: true,
    },
    mode: 'development'
};
```

6. Add a package.json file to download the webpack dependencies:

```
{
    "scripts": {
        "build": "webpack",
```

```
        "serve": "webpack-dev-server"
    },
    "devDependencies": {
        "html-webpack-plugin": "^5.5.0",
        "webpack": "^5.64.1",
        "webpack-cli": "^4.9.1",
        "webpack-dev-server": "^4.5.0"
    }
}
```

> **Note**
> Please use the latest version of the dependencies applicable here.

7. Create an index.js file to load the binding JavaScript that in turn loads the WebAssembly module generated:

```
import("./live_reload").then(module => {
    console.log(module.hello_world());
});
```

8. Now, head over to the terminal and install the npm dependencies using the following:

```
$ npm install
```

Run webpack-dev-server using the following:

```
$ npm run serve
```

We have already used webpack-dev-server to enable automatic recompiling. We can now go and change the HTML, CSS, or JavaScript file. Once we save the changes, the webpack server will compile everything. Once compiled, the changes are reflected in the browser.

But wait, what will happen if you change the Rust file? Let's try changing it:

```
use wasm_bindgen::prelude::*;


#[wasm_bindgen]
pub fn hello_world() -> String {
    "Hello Universe".to_string()
}
```

We made a huge change in our `main.rs` file. Yeah, we changed from *world* to *universe*; isn't that huge? But once you save the file, you will not see any changes in the browser. In fact, even the webpack compiler is not recompiling things.

The webpack compiler by default looks for the changes that will happen in the HTML, CSS, and JavaScript files (things that are defined in the configuration file and those that are included inside the dependency graph). But it has no idea about the Rust code.

We need to somehow tell webpack to look for the code changes in Rust. We can use a plugin for that, one that will look at any changes in the specified location of the specified file type. Then, it will retrigger the build process. We will use `wasm-pack-plugin` for this.

Add the `wasm-pack-plugin` dependency to the application using the following command:

```
$ npm i @wasm-tool/wasm-pack-plugin -D
```

Then, hook this plugin into webpack's plugin system via the `webpack.config.js` file:

```
const path = require('path');
const HtmlWebpackPlugin = require('html-webpack-plugin');
const WasmPackPlugin = require('@wasm-tool/wasm-pack-
  plugin');

module.exports = {
    entry: './index.js',
    output: {
        path: path.resolve(__dirname, 'dist'),
        filename: 'bundle.js',
    },
    plugins: [
        new HtmlWebpackPlugin(),
        new WasmPackPlugin({
                crateDirectory: path.resolve(__dirname)
        }),
    ],
    experiments: {
        syncWebAssembly: true,
    },
```

```
    mode: 'development'
};
```

We import `wasm-pack-plugin`. We specify the crate directory in which the `Cargo.toml` file is present and then the plugin will take care of the auto-reloading part. To see it in action, let's stop and start the webpack server using `npm run serve`.

Now, let's edit the `src/main.rs` file with Hello Galaxy. Open the browser to see the console log changed to **Hello Galaxy** already.

So, what happens here?

`wasm-pack-plugin` is hooked into webpack using webpack's plugin system. This will run along with the webpack compiler. If any changes are made in the `src` directory, `wasm-pack-plugin` will then run the `wasm-pack` compilation to compile the Rust code into WebAssembly modules automatically. This will trigger a recompilation in the webpack compiler. Once the webpack compiler recompiles, it will notify `webpack-dev-server` to reload the changes in the browser. The browser then reloads the changes automatically.

`wasm-pack-plugin` makes it easy to run Rust and WebAssembly along with webpack. Now, let's check how we can run Rust and WebAssembly with Parcel.

Bundling WebAssembly modules with Parcel

Parcel is a blazing-fast, zero-configuration web application bundler. Parcel is the new kid in the web application bundler space. It is built from scratch to be fast and needs zero configuration. The main pain point of webpack is its configuration. Although it looks simpler to start with, it gradually becomes more complex and unmanageable when the application grows. But the configuration will give a complete overview of what is happening and how it is bundling. With zero configuration, Parcel will infer the bundle from the initial point (that is, `index.html`) and then build the entire graph from there.

While webpack has a plugin-based architecture, Parcel has a worker-based architecture. This enables Parcel to be faster than webpack as it uses multicore compilation and cache.

Parcel also has inbuilt configuration to support JavaScript, CSS, and HTML files. Just like webpack, it also has various plugins that we can use to configure the bundler to produce the required output.

It also has inbuilt transformation support using standard Babel, PostCSS, and PostHTML when it is required. We can extend them and change them via plugins if needed.

Parcel also has automatic, out-of-the-box hot module replacement to track and record changes to files (that are recorded by the dependency graph). Let's build WebAssembly modules using parcel as a bundler:

1. We will start by creating a new Rust project:

```
$ cargo new --lib live_reload_parcel
  Created library `live_reload_parcel` package
```

Once the project is created, open the project in your favorite editor.

2. To define the wasm-bindgen dependency for the project, open the Cargo.toml file:

```
[package]
name = "live_reload_parcel"
version = "0.1.0"
authors = ["Sendil Kumar"]
edition = "2018"

[lib]
crate-type = ["cdylib"]

[dependencies]
wasm-bindgen = "0.2.38"
```

First, remove the [dependencies] tag and replace it with the bold lines above. We are telling the compiler that the library that is getting generated will be dynamic and it has a dependency on the wasm-bindgen library.

3. Then, we open the src/lib.rs file and replace the contents with the following:

```
use wasm_bindgen::prelude::*;

#[wasm_bindgen]
pub fn hello_world() -> String {
    "Hello World".to_string()
}
```

4. We will reuse the simple Hello World example here. Build the `wasm` module using the following:

```
$ cargo build --target wasm32-unknown-unknown
$ wasm-bindgen target/wasm32-unknown-
  unknown/debug/live_reload_parcel.wasm --out-dir .
```

5. Since Parcel supports zero configuration, all we need to do is add Parcel dependencies to `package.json`:

```
{
    "scripts": {
        "build": "parcel index.html",
        "serve": "parcel build index.html"
    },
    "devDependencies": {
        "parcel-bundler": "^1.12.3"
    }
}
```

Since Parcel is a zero-configuration bundler, we just have to define the entry point to it. We define the entry point in the `scripts` section. The `serve` command is the command that we use to run the code for development purposes. When we define `parcel build index.html`, we are informing Parcel that the entry point is `index.html`.

> **Note**
> Please use the latest version applicable here.

6. Then, we will create the entry point. We will create an `index.html` file as specified in the `package.json` script:

```
<html>
    <head>
        . . .
        <script src="./index.js"> </script>
    </head>
    <body> ... </body>
</html>
```

7. Create an `index.js` file to load the binding JavaScript, which in turn loads the WebAssembly module generated:

```
import { module } from './live_reload_parcel.js';
module.hello_world();
```

8. Now, head over to the terminal. Run the following command to install the dependencies:

```
$ npm install
```

Run the Parcel application using the following:

```
$ npm run serve
```

Parcel's zero-configuration nature makes it extremely easy to get started with WebAssembly. By default, Parcel supports `.wasm` files. We can even import `.wasm` files just like any other `.js` file.

It is important to note that synchronously importing WebAssembly modules is still not supported. But we can write an import as a synchronous import. Internally, Parcel will generate the necessary extra code to preload the file before JavaScript execution starts.

This implies the WebAssembly file will be a separate bundle rather than in line with the bundled JavaScript file.

9. Let's change the Rust file and see what happens:

```
use wasm_bindgen::prelude::*;


#[wasm_bindgen]
pub fn hello_world() -> String {
    "Hello Universe".to_string()
}
```

10. Once you save the file, you will not see any changes. Parcel has no clue that you have changed the source and the compiler will not react.

To make Parcel react to the Rust source changes, we need to add a plugin. The plugin is `parcel-plugin-wasm.rs`.

11. To install the plugin, we can run the following:

```
npm install -D parcel-plugin-wasm.rs
```

This will download the plugin to `node_modules`. This will also save the plugin in `package.json`'s `devDependencies`.

12. Once installed, we need to change `index.js` such that it looks at the source code directly instead of referencing from the `Cargo.toml` file:

```
import { hello_world } from './src/lib.rs';
// import { hello_world } from './Cargo.toml';
hello_world();
```

Here, instead of importing from the WebAssembly module, we specify the entry Rust file. We can even specify the location of the `Cargo.toml` file to make Parcel look for changes in the respective places.

13. Now, let's edit the `src/main.rs` file with Hello Galaxy. Open the browser to see how the console log has changed to **Hello Galaxy.**

So, what happens here?

Parcel just requires the starting point of our application. It will generate the dependency graph from there. The parcel plugin keeps looking for any changes in the folder to happen. It basically looks in the folder that contains the `Cargo.toml` file. The `Cargo.toml` location is given to the Parcel bundler and its plugin via `index.js`.

So, any changes that happen to the Rust file will lead to the following process.

When the Rust file is saved, the watchers inside `parcel-plugin-wasm.rs` are triggered. Then, `parcel-plugin-wasm.rs` will initiate the compilation process of the Rust code into WebAssembly via `wasm-pack`. Once `wasm-pack` compiles and produces new WebAssembly code, the plugin will notify the Parcel compiler that something in the dependency graph has been changed.

The Parcel compiler then recompiles, which will result in the browser being refreshed. The browser now displays the changed message.

Note that for Parcel, we actually used a synchronous module import, while for webpack, we were relying on asynchronous import.

The `parcel-plugin-wasm.rs` plugin makes it easy to run Rust and WebAssembly along with Parcel. Now, let's check how we can install and use `wasm-pack` to pack and publish WebAssembly modules.

Introducing wasm-pack

To be compatible with JavaScript, Rust-based WebAssembly applications should be completely interoperable with the JavaScript world. Without that, it will be difficult for developers to bootstrap their WebAssembly projects in JavaScript.

The node modules completely changed the perspective of the JavaScript world. They make it easier to develop and share the modules between Browser and Node environments. Developers around the world can use these libraries wherever and whenever they want.

> *The* wasm-pack *tool seeks to be a one-stop shop for building and working with Rust-generated WebAssembly that you would like to interoperate with JavaScript, in the browser or with Node.js. -* wasm-pack *website.*
> https://github.com/rustwasm/wasm-pack

Why do you need wasm-pack?

wasm-pack makes it easy to build and pack Rust- and WebAssembly-based projects. Once packed, the module is ready to be shared with the world via the npm registry – just like millions (or even billions) of JavaScript libraries out there.

How to use wasm-pack

wasm-pack is available as a Cargo library. If you are following along with this book, then you might have already installed Cargo. To install wasm-pack, run the following command:

```
$ cargo install wasm-pack
```

The preceding command will download, compile, and install the wasm-pack library. Once installed, the wasm-pack command will be available.

To check whether wasm-pack is installed correctly, run the following:

```
$ wasm-pack --version
wasm-pack 0.6.0
```

Once you have `wasm-pack` installed, let's see how to use `wasm-pack` to build and pack Rust and WebAssembly projects:

1. We will first generate a new project with Cargo. To generate the project, use the following:

```
$ cargo new --lib wasm_pack_world
  Created library `wasm_pack_world` package
```

Once the project is created, open it in your favorite editor.

2. To define the `wasm-bindgen` dependency for the project, open the `cargo.toml` file:

```
[package]
name = "wasm_pack_world"
version = "0.1.0"
authors = ["Sendil Kumar"]
edition = "2018"


[lib]
crate-type = ["cdylib"]


[dependencies]
wasm-bindgen = "0.2.38"
```

First, remove the `[dependencies]` tag and replace that with the wasm-bindgen library. We are telling the compiler that the library that is getting generated will be dynamic and it has a dependency to the `wasm-bindgen` library.

3. Then, we open the `src/lib.rs` file and replace the contents with the following:

```
use wasm_bindgen::prelude::*;


#[wasm_bindgen]
pub fn get_me_universe_answer() -> i32 {
    42
}
```

Again, this is a simple function that returns a number (which is the universal answer).

Previously, we used to build the Rust and WebAssembly application with `rustc` or Cargo. This produced a WebAssembly binary. But the binary is not useful by itself; it needs a binding file. With `wasm-bindgen`, we will generate the binding file along with the WebAssembly binary.

These two steps are mandatory, but they are mundane. We can replace them with `wasm-pack`.

4. To build the WebAssembly application with `wasm-pack`, run the following command:

```
$ wasm-pack build
```

When we run `wasm-pack build`, this is what happens:

I.   `wasm-pack` first checks whether the Rust compiler is installed. If it's installed, then it checks whether the Rust compiler is greater than version 1.30.

II. `wasm-pack` checks for the crate configuration and whether the library indicates that we are generating a dynamic library.

III. `wasm-pack` validates whether there is any `wasm-target` available for building. If the `wasm32-unknown-unknown` target is not available, `wasm-pack` will download and add the target.

IV. Once the environment is ready, `wasm-pack` then starts compiling the module and build the WebAssembly Module and binding JavaScript files.

Note that the `wasm-pack` command also generates the `package.json` file. The `package.json` file looks similar to this:

```json
{
"name": "wasm_pack_world",
"collaborators": [
"Sendil Kumar"
],
"version": "0.1.0",
"files": [
    "wasm_pack_world_bg.wasm",
    "wasm_pack_world.js",
    "wasm_pack_world.d.ts"
],
```

```
"module": "wasm_pack_world.js",
"types": "wasm_pack_world.d.ts",
"sideEffects": "false"
}
```

5. Finally, it copies over the Readme and LICENSE file if we have one, to ensure there is shared documentation between the Rust and WebAssembly versions.

    wasm-pack also checks for the presence of wasm-bindgen-cli, which, if not present, will be installed using Cargo.

6. When the build has successfully finished, it will create a pkg directory. Inside pkg, it will pipe the output of wasm-bindgen:

```
pkg
├── package.json
├── wasm_pack_world.d.ts
├── wasm_pack_world.js
├── wasm_pack_world_bg.d.ts
└── wasm_pack_world_bg.wasm
```

Now, this pkg folder can be bundled and shared like any other JavaScript module. We'll see how to achieve that in the future recipes.

wasm-pack is a great tool to pack and publish WebAssembly modules. Now, let's check out how to use it.

Packing and publishing using wasm-pack

The most amazing (and, of course, the most important) thing for a library developer to do is to pack and publish artifacts. That is why we spend our days and nights carefully crafting the application, publishing it to the world, receiving feedback (either negative or positive), and then enhancing the application based on that.

The critical point for any project is its first release, which defines the fate of the project. Even though it is simply an MVP, it will give the world a glimpse of what we are working on and gives us a glimpse of what we have to work on in the future.

`wasm-pack` helps us to build, pack, and publish Rust- and WebAssembly-based projects into the npm registry. We have already seen how `wasm-pack` makes it simpler to build Rust into the WebAssembly binary along with the binding JavaScript file using `wasm-bindgen` underneath. Let's further explore what we can do with its `pack` and `publish` flags.

`wasm-pack` provides a `pack` flag to pack the artifacts that were generated using the `wasm-pack` build command. Although it is not necessary to use `wasm-pack` to build binaries, it generates all the boilerplate that we will need to pack the artifacts into a Node module.

In order to pack the built artifacts using `wasm-pack`, we have to run the following command with reference to `pkg` (or the directory with which we generated our build artifacts):

```
$ wasm-pack pack pkg
```

We can also run the command by passing in `project_folder/pkg` as its argument. Without any argument, the `wasm-pack pack` command will search for the `pkg` directory in the current working directory where it is running.

The `wasm-pack pack` command first identifies whether the folder provided is a `pkg` directory or contains a `pkg` directory as its immediate child. If the check passes, then `wasm-pack` will invoke the npm pack command underneath, to pack the library into an npm package.

To bundle the npm package, all we need is a valid `package.json` file. That file is generated by the `wasm-pack` build command.

We can run the `pack` command inside the `cg-array-world` example from our previous recipe and check what happens:

```
$ wasm-pack pack
npm notice
npm notice 📦 cg-array-world@0.1.0
npm notice === Tarball Contents ===
npm notice 313B package.json
npm notice 32.7kB cg_array_world_bg.wasm
npm notice 135B cg_array_world.d.ts
npm notice 1.6kB cg_array_world.js
npm notice 1.5kB README.md
npm notice === Tarball Details ===
```

```
npm notice name: cg-array-world
npm notice version: 0.1.0
npm notice filename: cg-array-world-0.1.0.tgz
npm notice package size: 16.0 kB
npm notice unpacked size: 36.4 kB
npm notice shasum: 243488f1f5a859b60bb34f39146b35ba720dd8ea
npm notice integrity: sha512-9SFuObzEpi254[...]/1kHq6RKSgnNw==
npm notice total files: 5
npm notice
cg-array-world-0.1.0.tgz
 | 🎒 packed up your package!
```

As you can see here, the `pack` command creates a tarball package with the contents inside the `pkg` folder with the help of the `npm pack` command.

Once we have packed our application, the obvious next step will be to publish it. In order to publish the tarball generated, `wasm-pack` has a `publish` option.

In order to publish the package, we have to run the following command:

```
$ wasm-pack publish
```

The `wasm-pack publish` command will first check whether the `pkg` directory is already present in the directory provided.

If the `pkg` directory is not present, then it will ask whether you want to create the package first:

```
$ wasm-pack publish
Your package hasn't been built, build it? [Y/n]
```

If you answer `Y` to the question, then it asks for you to input the folder in which you want to generate the build artifacts. We can give any folder name or use the default:

```
$ wasm-pack publish
Your package hasn't been built, build it? yes
out_dir[default: pkg]:
```

Then, it asks for your target, that is, the target in which the build should be generated. You can choose between the various options here, as discussed in the build recipe:

```
$ wasm-pack publish
Your package hasn't been built, build it? yes
out_dir[default: pkg]: .
target[default: browser]:
> browser
nodejs
no-modules
```

Based on the option provided, it will generate the artifact in the specified folder.

Once the artifacts are produced, they are then ready to be published using npm publish. For npm publish to work correctly, we need to be authenticated. You can authenticate to npm by using either npm login or wasm-pack login.

The wasm-pack login command will invoke the underlying npm login command and then create a session:

```
$ wasm-pack login
Username: sendilkumarn
Password: *************
login succeeded.
```

The wasm-pack publish command also supports two options, namely the following:

- -a or --access to determine the access level of the package to be deployed.

  This accepts either public or restricted:
  - public – Makes the package public
  - restricted – Makes the package internal
- -t or --target to support various targets in which the build is produced.

Thus, wasm-pack makes it easy to pack and publish WebAssembly binaries.

Summary

In this chapter, we saw how to run a WebAssembly project with bundlers such as webpack and Parcel. Parcel and webpack make it easy for JavaScript developers to run and develop Rust and WebAssembly projects. Then, we installed `wasm-pack` and used it to run the project. Finally, we used `wasm-pack` to pack and publish the WebAssembly module to npm.

In the next chapter, we will explore how to share complex objects between Rust and WebAssembly with `wasm-bindgen`.

9
Crossing the Boundary between Rust and WebAssembly

So far, we have only seen examples of sharing simple numbers between JavaScript and WebAssembly. In the last section, we saw how `wasm-bindgen` helps to pass a string from Rust to JavaScript with ease. In this chapter, we will explore how `wasm-bindgen` makes it easier to convert more complex data types between JavaScript and WebAssembly via Rust. We will cover the following sections in this chapter:

- Sharing classes from Rust with JavaScript
- Sharing classes from JavaScript with Rust
- Calling the JavaScript API via WebAssembly
- Calling closures via WebAssembly
- Importing the JavaScript function into Rust
- Calling a web API via WebAssembly

Technical requirements

You can find the code files present in this chapter on GitHub at `https://github.com/PacktPublishing/Practical-WebAssembly/tree/main/09-rust-wasm-boundary`.

Sharing classes from Rust with JavaScript

`wasm-bindgen` enables sharing classes from JavaScript with Rust and vice versa using simple annotations. It handles all the boilerplate stuff, such as translating a value from JavaScript to WebAssembly or WebAssembly to JavaScript, complex memory manipulations, and error-prone pointer arithmetic. Thus, `wasm-bindgen` makes everything easier.

Let's see how easy it is to share classes between JavaScript and WebAssembly (from Rust):

1. Create a new project:

```
$ cargo new --lib class_world
Created library `class_world` package
```

2. Define the `wasm-bindgen` dependency for the project. Open the `cargo.toml` file and add the following content:

```
[package]
name = "class_world"
version = "0.1.0"
authors = ["Sendil Kumar"]
edition = "2018"
[lib]
crate-type = ["cdylib"]
[dependencies]
wasm-bindgen = "0.2.68"
```

3. Open the `src/lib.rs` file and replace the content with the following:

```
use wasm_bindgen::prelude::*;
#[wasm_bindgen]
pub struct Point {
    x: i32,
    y: i32,
}
```

```rust
#[wasm_bindgen]
impl Point {
    pub fn new(x: i32, y: i32) -> Point {
        Point { x: x, y: y}
    }
    pub fn get_x(&self) -> i32 {
        self.x
    }
    pub fn get_y(&self) -> i32 {
        self.y
    }

    pub fn set_x(&mut self, x: i32) {
        self.x = x;
    }

    pub fn set_y(&mut self, y:i32) {
        self.y = y;
    }

    pub fn add(&mut self, p: Point) {
        self.x = self.x + p.x;
        self.y = self.y + p.y;
    }
}
```

> **Note**
>
> The &mut before the argument specifies that the argument (in this case, self) is a mutable reference.

Rust does not have classes but we can define a class via a struct. The Point struct contains getters, setters, and an add function. This is normal Rust code with only the #[wasm_bindgen] annotation added.

> **Note**
>
> The functions and struct were explicitly marked pub. The pub modifier means the function is public and will be exported.

4. Generate the WebAssembly module using Cargo:

    ```
    $ cargo build --target=wasm32-unknown-unknown
    ```

5. Use the `wasm-bindgen` CLI to generate the binding file for the WebAssembly module generated:

    ```
    $ wasm-bindgen target/wasm32-unknown-
        unknown/debug/class_world.wasm --out-dir .
    ```

    This will generate the binding files and type definition files as we have seen in the previous chapter. Let's look at the `class_world.js` file first. This file will be similar to the file generated in previous chapters except for the `Point` class. The `Point` class has all the getters, the setters, and the `add` function in it. The functions use the pointer to their references.

Additionally, `wasm-bindgen` produces a static method called `__wrap` that creates the `Point` class object and attaches a pointer to it. It adds a free method that in turn calls the `__wbg_point_free` method inside the WebAssembly module. This method is responsible for freeing up the memory taken by the `Point` object or class.

Create the following files. We will use them in the other sections too:

1. Create `webpack.config.js`. This holds the webpack configuration:

    ```
    const path = require('path');
    const HtmlWebpackPlugin = require('html-webpack-
        plugin');
    module.exports = {
        entry: './index.js',
        output: {
            path: path.resolve(__dirname, 'dist'),
            filename: 'bundle.js',
        },
        plugins: [
            new HtmlWebpackPlugin(),
        ],
        mode: 'development'
    };
    ```

2. Create `package.json` and add the following content:

```
{
    "scripts": {
        "build": "webpack",
        "serve": "webpack-dev-server"
    },
    "dependencies": {
        "html-webpack-plugin": "^3.2.0",
        "webpack": "^4.41.5",
        "webpack-cli": "^3.3.10",
        "webpack-dev-server": "^3.10.1"
    }
}
```

3. Create an `index.js` file:

```
$ touch index.js
```

4. Then, run `npm install`. Modify `index.js` with the following content:

```
import("./class_world").then(({Point}) => {
const p1 = Point.new(10, 10);
console.log(p1.get_x(), p1.get_y());
const p2 = Point.new(3, 3);
p1.add(p2);
console.log(p1.get_x(), p1.get_y());
});
```

We call the new method in the `Point` class and pass it x and y. We print the x and y coordinates. This will print 10, 10. Then, we will create another point (p2). Finally, we call the `add` function and pass it point p2. This will print 13, 13.

5. The getter method uses the pointer and fetches the value from the shared array:

```
get_x() {
    return wasm.point_get_x(this.ptr);
}
```

6. In the setter method, we pass in the pointer and the value. Since we are just passing in a number here, there is no extra conversion needed:

```
set_x(arg0) {
    return wasm.point_set_x(this.ptr, arg0);
}
```

7. In the case of add, we take the argument, get the pointer to the Point object, and pass it to the WebAssembly module:

```
add(arg0) {
    const ptr0 = arg0.ptr;
    arg0.ptr = 0;
    return wasm.point_add(this.ptr, ptr0);
}
```

wasm-bindgen makes it simple to convert a class into a WebAssembly module. We have seen how to share a class in Rust with JavaScript. Now, we will see how to share a class from JavaScript with Rust.

Sharing classes from JavaScript with Rust

Sharing JavaScript classes with Rust is also made easy with #[wasm_bindgen]. Let's look at how to achieve it.

JavaScript classes are objects with some methods. Rust is a strictly typed language. This means the Rust compiler needs to have concrete bindings. Without them, the compiler complains, because it needs to know about the lifetime of an object. We need a way to ensure the compiler has this API available at runtime.

The extern C function block helps out here. The extern C makes a function name available in Rust.

In this example, let's see how to share a class from JavaScript with Rust:

1. Let's create a new project:

```
$ cargo new --lib class_from_js_world
Created library `class_from_js_world` package
```

2. Define the `wasm-bindgen` dependency for the project. Open the `cargo.toml` file and add the following content:

```
[package]
name = "class_from_js_world"
version = "0.1.0"
authors = ["Sendil Kumar"]
edition = "2018"

[lib]
crate-type = ["cdylib"]

[dependencies]
wasm-bindgen = "0.2.68"
```

Please copy over `package.json`, `index.js`, and `webpack-config.js` from the previous section. Then, run `npm install`.

3. Open the `src/lib.rs` file and replace the contents with the following:

```
use wasm_bindgen::prelude::*;

#[wasm_bindgen(module = "./point")] . // 1
extern "C" {
    pub type Point; // 2

    #[wasm_bindgen(constructor)] //3
    fn new(x: i32, y: i32) -> Point;

    #[wasm_bindgen(method, getter)] //4
    fn get_x(this: &Point) -> i32;

    #[wasm_bindgen(method, getter)]
    fn get_y(this: &Point) -> i32;

    #[wasm_bindgen(method, setter)] //5
    fn set_x(this: &Point, x:i32) -> i32;

    #[wasm_bindgen(method, setter)]
```

```
    fn set_y(this: &Point, y:i32) -> i32;

    #[wasm_bindgen(method)] // 6
    fn add(this: &Point, p: Point);
}

#[wasm_bindgen]
fn get_precious_point() -> Point { //7
    let p = Point::new(10, 10);
    let p1 = Point::new(3, 3);
    p.add(p1); // 8
    p
}
```

At //1, we are importing the JavaScript module. This will import a JavaScript file, point.js. Note that this file should be present in the same directory as Cargo. toml. Then, we create an extern C block to define the methods that we need to use.

We first declare a type in the block (pub type Point;). Now, we can use this as any other type in the Rust code. After that, we define a list of functions. We first define the constructor. We pass in the constructor as an argument to the #[wasm_ bindgen] annotation. Define a function that takes in arguments and returns the type declared previously. This will bind to the namespace of the Point type, and we can call Point::new(x, y); inside the Rust function.

Then, we define getters and setters (//4 and //5, respectively). We can even define a method; these are analogous to the function on the JavaScript side. Then, we have the add function.

Note

All the functions inside the extern C block are completely typed.

Finally, we are exporting the get_precious_point() function using the #[wasm_bindgen] annotation. In the get_precious_point function, we create two Point with Point::new(x, y), then add two points using p1.add(p2).

We can call this from JavaScript just like we did before. We also have to define a Point class on the JavaScript side.

4. Create `Point.js` with the following content:

```
export class Point {
    constructor(x, y) {
        this.x = x;
        this.y = y;
    }

    get_x() {
        return this.x;
    }

    get_y() {
        return this.y;
    }

    set_x(x) {
        this.x = x;
    }

    set_y(y) {
        this.y = y;
    }

    add(p1) {
        this.x += p1.x;
        this.y += p1.y;
    }
}
```

5. Finally, replace `index.js` with the following:

```
import("./class_from_js_world").then(module => {
    console.log(module.get_precious_point());
});
```

6. Now, run the following command to start the server:

```
$ npm run serve
```

7. Open the browser and run `http://localhost:8000`. Open the developer console to see the printed object class.

8. Let's see how the `#[wasm_bindgen]` macro is expanding the code:

```
$ cargo expand --target=wasm32-unknown-unknown >
  expanded.rs
```

There are a few interesting things happening here.

First, the `type` point is converted into a struct. This is similar to what we did in the previous example. But the struct's members are JsValue instead of x and y. This is because `wasm_bindgen` will not know what this Point class is instantiating. So, it creates a JavaScript object and makes that its member:

```
pub struct Point {
    obj: ::wasm_bindgen::JsValue,
}
```

It also defines how to construct the Point object and how to dereference it. This is useful for the WebAssembly runtime to know when to allocate and when to dereference it.

All the methods that are defined are converted into the implementation of the Point struct. As you can see, there is a lot of unsafe code in the method declaration. This is because the Rust code interacts directly with the raw pointers:

```
fn new(x: i32, y: i32) -> Point {
#[link(wasm_import_module =
  "__wbindgen_placeholder__")]
extern "C" {
fn __wbg_new_3ffc5ccd013f4db7(x:<i32 as
  ::wasm_bindgen::convert::IntoWasmAbi>::Abi, y:<i32 as
  ::wasm_bindgen::convert::IntoWasmAbi>::Abi) -> <Point
  as ::wasm_bindgen::convert::FromWasmAbi>::Abi;
}

unsafe {
let _ret = {
let mut __stack =
  ::wasm_bindgen::convert::GlobalStack::new();
let x = <i32 as
```

```
    ::wasm_bindgen::convert::IntoWasmAbi>::into_abi
    (x, &mut __stack);
  let y = <i32 as
    ::wasm_bindgen::convert::IntoWasmAbi>::into_abi
    (y, &mut __stack);
  __wbg_new_3ffc5ccd013f4db7(x, y)
};
<Point as
  ::wasm_bindgen::convert::FromWasmAbi>::from_abi(_ret,
  &mut ::wasm_bindgen::convert::GlobalStack::new())
}
}
```

Shown in the previous code is the code generated by the #[wasm_bindgen(constructor)] macro. It first links the code with the extern C block. The arguments are then cast such that they are inferred in WebAssembly.

Then, we have the unsafe block. First, space is reserved in the global stack. Then, both x and y are converted into the IntoWasmAbi type.

IntoWasmAbi is a trait for anything that can be converted into a type that can cross the WebAssembly ABI directly, for example, u32 or f64. Then, the function in JavaScript is called. The returned value is then cast into a Point type using FromWasmAbi.

FromWasmAbi is a trait for anything that can be recovered by value from the WebAssembly ABI boundary; for example, a Rust u8 can be recovered from the WebAssembly ABI u32 type.

We have seen how to share a class in JavaScript with Rust. Now, we will see how we can call a JavaScript API in Rust.

Calling the JavaScript API via WebAssembly

JavaScript provides a rich API to work with objects, arrays, maps, sets, and so on. If we want to use or define them in Rust, then we need to provide the necessary bindings. Handcrafting those bindings will be a huge process. But what if we have bindings to those APIs? That is a common API for both Node.js and a browser environment that will create a platform where we can write the code completely in Rust and use wasm_bindgen to create necessary code.

The rustwasm team's answer to that is the js-sys crate.

> *The js-sys crate contains raw* #[wasm_bindgen] *bindings to all the global APIs guaranteed to exist in every JavaScript environment by the ECMAScript standard. – RustWASM*

They provide bindings to JavaScript's standard built-in objects, including their methods and properties.

In this example, let's see how to call a JavaScript API via WebAssembly:

1. Create a default project with the cargo new command:

    ```
    $ cargo new --lib jsapi
    ```

2. Copy over webpack.config.js, index.js, and package.json similarly to the previous example. Then, open the generated project in your favorite editor.

3. Change the contents of Cargo.toml:

    ```
    [package]
    name = "jsapi"
    version = "0.1.0"
    authors = ["Sendil Kumar"]
    edition = "2018"

    [lib]
    crate-type = ["cdylib"]

    [dependencies]
    wasm-bindgen = "0.2.68"
    js-sys = "0.3.45"
    ```

4. Now, open `src/lib.rs` and replace the file with the following content. We can create a JavaScript map inside Rust using the following snippet:

```
use wasm_bindgen::prelude::*;

use js_sys::Map;

#[wasm_bindgen]
pub fn new_js_map() -> Map {
    Map::new()
}
```

Added to the `wasm_bindgen` import, we imported the map from the `js_sys` crate using `use js_sys::Map;`.

5. Then, we define the `new_js_map` function, which will return a new map:

```
#[wasm_bindgen]
pub fn set_get_js_map() -> JsValue {
    let map = Map::new();
    map.set(&"foo".into(), &"bar".into());
    map.get(&"foo".into())
}
```

The `set_get_js_map` function creates a new map, sets a value in the map, and then returns the value set.

Note that the return type here is `JsValue`. This is a wrapper in Rust for specifying the JavaScript values. Also, note that we are passing the string into the trait functions get and set. This will return `bar` as the output when called in JavaScript.

6. We now also run through the map using `for_each` inside the Rust code like this:

```
#[wasm_bindgen]
pub fn run_through_map() -> f64 {
    let map = Map::new();
    map.set(&1.into(), &1.into());
    map.set(&2.into(), &2.into());
    map.set(&3.into(), &3.into());
    map.set(&4.into(), &4.into());
    map.set(&5.into(), &5.into());
    let mut res: f64 = 0.0;
```

```
map.for_each(&mut |value, _| {
    res = res + value.as_f64().unwrap();
});

res
}
```

This creates a map and then loads the map with the values 1, 2, 3, 4, and 5. Then, it runs over the created map and adds the value together. This will produce an output of 15 (that is, $1 + 2 + 3 + 4 + 5$).

7. Lastly, we replace index.js with the following content:

```
import("./jsapi").then(module => {
    let m = module.new_js_map();
    m.set("Hi", "Hi");
    console.log(m); // prints Map { "Hi" -> "Hi" }
    console.log(module.set_get_js_map());  // prints
      "bar"
    console.log(module.run_through_map()); // prints
      15
});
```

Running this on the browser will print the result. Refer to the comments near the console log statements.

Let's start with the generated JavaScript binding file. The generated binding JavaScript file has almost the same structure as in the previous section, but with a few more functions exported.

The heap object is used as a stack here. All the JavaScript objects that are shared or referenced with the WebAssembly modules are stored in this heap. It is also important to note that once a value is accessed, it is popped out from the heap.

```
function takeObject(idx) {
    const ret = getObject(idx);
    dropObject(idx);
    return ret;
}
```

The `takeObject` function is used to fetch the object from the heap. It first gets the object at the given index. Then, it removes the object from that heap index (that is, it pops it out). Finally, it returns the value `ret`.

Similarly, we can use JavaScript APIs inside Rust. The bindings are only generated for the common JavaScript API (including Node.js and the browser).

We have seen how to call the JavaScript API in Rust. Now, we will see how we can call a Rust closure via WebAssembly.

Calling closures via WebAssembly

The official Rust book defines closures as follows:

> *Closures are anonymous functions which you can save in a variable or can be passed as arguments to other functions. - The Rust Programming Language (Covers Rust 2018) by Steve Klabnik and Carol Nichols (*`https://doc.rust-lang.org/book/ch13-00-functional-features.html`*)*

MDN defines a closure for JavaScript as follows:

> *A closure is the combination of a function and lexical environment within which that function was declared.- MDN Web Docs (*`https://developer.mozilla.org/en-US/docs/Web/JavaScript/Closures#closure`*)*

In general, closures are self-contained blocks of functionality that are tossed around and used in the code. They can capture and store references to the variables from the context in which they are defined.

Closures and functions are similar except for a subtle difference. Closures will capture the state when it is first created. Then, whenever a closure is called, it closes over that captured state.

Closures are functions with a state. When you create a closure, it captures the state. Then, we can pass around closures just like any other function. When a closure is then invoked, it closes over this captured state and executes (even when the closure is invoked outside of their captured state). That is one of the important reasons why the use of closures is increasing on the functional side of JavaScript.

Closures make it easy to do data encapsulation, higher-order functions, and memoization. (Sounds functional, right? ;))

Let's see how to share closures from JavaScript to Rust and vice versa:

1. Create a new project:

```
$ cargo new --lib closure_world
        Created library `closure_world` package
```

2. Define the wasm-bindgen dependency for the project. Let's open the cargo.toml file and add the content highlighted in bold:

```
[package]
name = "closure_world"
version = "0.1.0"
authors = ["Sendil Kumar"]
edition = "2018"


[lib]
crate-type = ["cdylib"]


[dependencies]
wasm-bindgen = "0.2.38"
js-sys = "0.3.15"
```

We will need the js-sys crate to copy over the closures from JavaScript into Rust. Please copy over package.json, index.js, and webpack-config.js from the previous section. Then, run npm install.

3. We then open the src/lib.rs file and add the content from our Point class example with an additional method that takes in JavaScript's closure function as its argument:

```
use wasm_bindgen::prelude::*;


#[wasm_bindgen]
pub struct Point {
    x: i32,
    y: i32,
}
```

```rust
#[wasm_bindgen]
impl Point {
    pub fn new(x: i32, y: i32) -> Point {
        Point { x: x, y: y}
    }

    pub fn get_x(&self) -> i32 {
        self.x
    }

    pub fn get_y(&self) -> i32 {
        self.y
    }

    pub fn set_x(&mut self, x: i32) {
        self.x = x;
    }

    pub fn set_y(&mut self, y:i32) {
        self.y = y;
    }

    pub fn add(&mut self, p: Point) {
        self.x = self.x + p.x;
        self.y = self.y + p.y;
    }

    pub fn distance(&self, js_func: js_sys::Function)
      -> JsValue {
        let this = JsValue::NULL;
        let x = JsValue::from(self.x);
        let y = JsValue::from(self.y);
        js_func.call2(&this, &x, &y).unwrap()
    }
}
```

Now, we will change `index.js` to call the `distance` function with a closure:

```
import("./closure_world").then(({Point}) => {
    const p1 = Point.new(13, 10);
    console.log(p1.distance((x, y) => x - y));
});
```

Let's spin the webpack server with `npm run serve`. This will print out 3.

The js-sys crate offers an option to invoke the JavaScript function using the apply and call method. That is what we have done by calling `js_func.call2(&this, &x, &y)`.

Rust does not have function overloading. This means that we have to use different method names based on the number of arguments that we pass. So, `js-sys` offers us `call1`, `call2`, `call3`, and so on, each taking 1, 2, 3, and so on arguments, respectively.

Invoking a JavaScript function in Rust will return `Result<JsValue, Error>`. We will unwrap the result to get the JsValue and return it. `wasm-bindgen` will create the necessary binding to return the value as a number in JavaScript.

On the other hand, passing a closure from Rust to JavaScript will need some extra information and options.

`wasm-bindgen` supports two variants here:

- Stack lifetime closures
- Heap-allocated closures

Let's see what they actually mean:

- Stack lifetime closures should not be invoked by JavaScript again once the imported JavaScript function that the closure was passed to returns. This is because once the function (closure) returns, the closure will be invalidated by Rust. Any future invocations will lead to an exception. In other words, stack lifetime closures are short-lived and they go out of context once they are accessed.

- On the other hand, heap-allocated closures are useful for invoking the memory multiple times. Here, the validity is tied to the lifetime of the closure in Rust. Once the closure in Rust is dropped, the closure will deallocate and garbage will be collected. This will in turn invalidate the closure (function) in JavaScript. Once invalidated, any further attempts to access the closure or memory will raise an exception.

Both the stack lifetime and heap-allocated closures support both `Fn` and `FnMut` closures, arguments, and return values.

We have seen how to call a closure function. Now, we will see how to import a function from JavaScript into Rust.

Importing the JavaScript function into Rust

In certain places, JavaScript is faster than WebAssembly because there is no overhead of boundary crossing and instantiating a separate runtime environment. JavaScript runs more naturally in its own environment.

The JavaScript ecosystem is huge. There are millions of libraries created and battle tested (not all of them, of course) with JavaScript. This makes JavaScript easy (easy here is subjective).

WebAssembly addresses the most important problem that we have in the frontend world, that of "consistent" performance. But it is not a complete replacement for JavaScript. WebAssembly helps JavaScript to deliver better and more consistent performance.

JavaScript will be a default choice in most places. It is important to provide an ecosystem that allows seamless integration between the two. We have already seen how to import a class from JavaScript into Rust. Similarly, we can import anything from JavaScript into Rust using wasm-bindgen. The most important part is that we can more naturally use these imported JavaScript functions inside Rust code.

In this example, let's see how to import a JavaScript function into Rust:

1. Create a new project:

```
$ cargo new --lib import_js_world
        Created library `import_js_world` package
```

2. Define the wasm-bindgen dependency for the project. Let's open the cargo.toml file and add the content highlighted in bold:

```
[package]
name = "import_js_world"
version = "0.1.0"
authors = ["Sendil Kumar"]
edition = "2018"


[lib]
crate-type = ["cdylib"]
```

```
[dependencies]
wasm-bindgen = "0.2.38"
```

3. Please copy over package.json, index.js, and webpack-config.js from the previous section. Then, run npm install. Then, open the src/lib.rs file and replace its content with the following:

```rust
use wasm_bindgen::prelude::*;

#[wasm_bindgen(module = "./array")]
extern "C" {
    fn topArray() -> f64;
    fn getNumber() -> i32;
    fn lowerCase(str: &str) -> String;
}

#[wasm_bindgen]
pub fn sum_of_square_root() -> f64 {
    let n = getNumber();
    let mut sum = 0;

    for _ in 0..n {
        sum = sum + (topArray().sqrt() as i64);
    }
    sum
}

#[wasm_bindgen]
pub fn some_string_to_share() -> String {
    lowerCase("HEYA! I AM ALL CAPS")
}
```

We start by importing the `wasm_bindgen` library. Then, we define the extern C block to define the FFI functions (that is, the functions that we import from JavaScript). Inside the extern C block, we define the function signature similar to what the Rust compiler understands. We also annotate the extern C block with `#[wasm_bindgen(module ="./array")]`. This helps the `wasm-bindgen` CLI to understand where the functions are defined and exported. It will use this information and create the necessary link.

4. The `array.js` file is in the same directory as the `cargo.toml` file. We will define `array.js` as follows:

    ```javascript
    let someGlobalArray = [1, 4, 9, 16, 25];

    export function getNumber() {
        return someGlobalArray.length;
    }

    export function topArray() {
        return someGlobalArray.sort().pop();
    }

    export lowerCase(str) {
        return str.toLowerCase();
    }
    ```

    The functions mentioned previously should be exported in the JavaScript file.

    We then declare a function (`sum_of_square_root`) in Rust and export it as a function in the WebAssembly module that is generated. We first call the `getNumber()` method from JavaScript. We use the return value and then run the `for` loop for the length of the array. For each loop, we call `topArray` to get the lowest element from the array. Then, we take the square root of the number (this happens in Rust code). Sum them up and return the sum (`15` for the example that we saw previously).

5. We will replace `index.js` with the following content:

    ```javascript
    import("./import_js_world").then(module => {
        console.log(module.sum_of_square_root());
        console.log(module.some_string_to_share());
    });
    ```

6. Let's run the previous code with `npm run serve`. Go to the URL and open the developer console. `15` and `HEYA! I AM ALL CAPS` will be printed in the console.

7. Open the generated binding JavaScript file. It will be interesting to see that both the `getNumber` and `topArray` functions are not available in the generated binding JavaScript file. The main reason for this is we are just sharing numbers between the JavaScript and the WebAssembly module. Hence, the boundary crossing happens more naturally in this case.

    But there will be a function exported for the `some_string_to_share` function. This is because we are sharing a string, which needs conversion. So, the binding file will make the necessary conversion to work with the string. It is also important to note that there is a `getInt64Memory` function. This is because we are returning `f64` as output. `wasm-bindgen` converts the number into `bigInt` and returns it to JavaScript.

8. `wasm-bindgen` also does the necessary shifting and parsing of the byte buffer based on the byte size of the memory object. For `Uint32Array`, the pointer and memory are calculated as follows:

    ```
    const rustptr = mem[retptr / 4];
    const rustlen = mem[retptr / 4 + 1];
    ```

9. For `BigInt64Array`, the pointer and memory are calculated as follows:

    ```
    const rustptr = mem[retptr / 8];
    const rustlen = mem[retptr / 8 + 1];
    ```

We have seen how to import a JavaScript function into Rust. Now, we will see how we can call a web API in Rust.

Calling a web API via WebAssembly

The evolution of the web has been phenomenal, with its growth being attributed to its open standards. Today, the web provides hundreds of APIs, which makes it easy for web developers to develop for audio, video, canvases, SVGs, USBs, batteries, and so on.

The web is universal and omnipresent. It is continuously experimented with and changed to make it desirable and easy for developers and companies to use, respectively. The web-sys crate provides access to almost all the APIs that are available on the web at the moment.

> *The* web-sys *crate provides raw bindings to all the Web's APIs: everything from DOM manipulation to WebGL to Web Audio to timers to fetch and more! – web-sys crates.io (*https://crates.io/crates/web-sys*)*

The WebIDL interface definitions are converted into wasm-bindgen's internal **abstract syntax trees** (**ASTs**). Then, these ASTs are used to create zero-overhead Rust and JavaScript glue code.

With the help of this binding code, we can call and manipulate the web APIs. Converting the web APIs into Rust ensures the type information of parameters and return values is handled correctly and safely.

In this example, let's call a web API via WebAssembly:

1. Create a default project with the cargo new command:

```
$ cargo new --lib web_sys_api
    Created library `web_sys_api` package
```

2. Copy over webpack.config.js, index.js, and package.json similarly to the jsapi section (in the above section). We'll now open the generated project in our favorite editor. Let's change the contents of cargo.toml:

```
[package]
name = "web_sys_api"
version = "0.1.0"
authors = ["Sendil Kumar"]
edition = "2018"

[lib]
crate-type = ["cdylib"]

[dependencies]
wasm-bindgen = "0.2.38"

[dependencies.web-sys]
version = "0.3.4"
```

```
features = [
    'Document',
    'Element',
    'HtmlElement',
    'Node',
    'Window',
]
```

The major difference here is that instead of just defining the dependency and its version, we also defined the features that we will be using in this example.

Why do we need it? Since there is a huge number of APIs in the web ecosystem, we do not want to carry bindings for all of them. The binding files are created and used only for the listed features.

3. Let's open `src/lib.rs` and replace the file with the following content:

```
use wasm_bindgen::prelude::*;

#[wasm_bindgen]
pub fn draw(percent: i32) -> Result<web_sys::Element,
  JsValue> {
    let window = web_sys::window().unwrap();
    let document = window.document().unwrap();

    let div = document.create_element("div")?;
    let ns = Some("http://www.w3.org/2000/svg");

    div.set_attribute("class", "pie")?;

    let svg = document.create_element_ns( ns, "svg")?;
    svg.set_attribute("height", "100")?;
    svg.set_attribute("width", "100")?;
    svg.set_attribute("viewBox", "0 0 32 32")?;

    let circle = document.create_element_ns(ns,
      "circle")?;
    circle.set_attribute("r", "16")?;
    circle.set_attribute("cx", "16")?;
```

```
circle.set_attribute("cy", "16")?;
circle.set_attribute("stroke-dasharray",
  &(percent.to_string().to_owned() +" 100"))?;


svg.append_child(&circle)?;


div.append_child(&svg)?;


Ok(div)
}
```

We first fetch the window using web_sys::window(). The unwrap at the end makes sure the window is available. If it is not, it will throw an error. After that, we get the document from the window object. We then create a div element with document.createElement. Then, we create an SVG and circle document element and append the circle to the SVG element. Finally, we append the SVG as a child to the div element and return the div element.

The API is quite similar to the web API except for the fact that the method names are using snake case instead of camel case.

4. We will change index.js to use this element as a web component:

```
import("./web_sys_api").then(module => {
    class Pie extends HTMLElement {
        constructor() {
            super();
            let shadow = this.attachShadow({ mode:
              'open' });
            let style =
              document.createElement('style');

            style.textContent = `
                svg {
                    width:100px;
                    height: 100px;
                    background: yellowgreen;
                    border-radius: 50%;
                }
```

```
            circle {
                fill: yellowgreen;
                stroke: #655;
                stroke-width: 32;
            }`;

        shadow.appendChild(module.draw(this.
        getAttribute
        ('value'));
        shadow.appendChild(style);
    }
}

customElements.define('pie-chart', Pie);

setInterval(() => {
    let r = Math.floor(Math.random() * 100);
    document.getElementsByTagName('body')[0].
    innerHTML = `
        <pie-chart value='${r}' />`;
}, 1000);
});
```

So, what have we done here? We first imported the binding file, which will, in turn, initialize the WebAssembly module. Once the WebAssembly module is initialized, we create a `Pie` class that extends the HTML element. Inside the class' constructor, we call the `super` method. Then, we create a shadow DOM. We add a style element to the shadow DOM, and then define the style for the element.

We go on to append the style element to the shadow element, and then add the element that is exported from the Rust code. We then register it as a custom element named `pie-chart`. Finally, we append the custom element to our document's body to see the pie chart getting displayed.

5. Now, run the following command:

```
$ npm run serve
```

Open the browser to see the pie chart.

Summary

In this chapter, we saw how `wasm-bindgen` makes it easy to share complex objects between JavaScript and Rust. The annotations make it easy to mark a function to export/import between JavaScript and WebAssembly. We also saw how js-sys and web-sys Cargo make it easier to call JavaScript and web APIs inside Rust code easily.

In the next chapter, we will see how to optimize the generated WebAssembly module in Rust.

10
Optimizing Rust and WebAssembly

So far, we have seen how Rust makes it easy to create and run WebAssembly modules and various tools provided by the Rust community. We will cover the following sections in this chapter:

- Minimizing the WebAssembly modules
- Analyzing the memory model in the WebAssembly module
- Analyzing the WebAssembly module with Twiggy

Technical requirements

You can find the code files present in this chapter on GitHub at `https://github.com/PacktPublishing/Practical-WebAssembly`.

Minimizing the WebAssembly modules

wasm-bindgen is a complete suite that generates the binding JavaScript file (along with polyfills) for the WebAssembly module. In previous chapters, we saw how wasm-bindgen provides libraries and makes it easy to pass complex objects between JavaScript and WebAssembly. But in the WebAssembly world, it is important to optimize the generated binary for size and performance.

Let's see how we can further optimize the WebAssembly modules:

1. Create a WebAssembly application with all the necessary toolchains:

    ```
    $ npm init rust-webpack wasm-rust
    🦀 Rust + 🕸 WebAssembly + Webpack = ♥
    ```

    This previous command creates a new Rust and JavaScript-based application with webpack as the bundler.

2. Go into the generated wasm-rust directory:

    ```
    cd wasm-rust
    ```

    The Rust source files are present in the src directory and the JavaScript files are available in the js directory. We have webpack configured for running the application.

3. Remove all the code from src/lib.rs and replace it with the following:

    ```rust
    use wasm_bindgen::prelude::*;


    #[cfg(feature = "wee_alloc")]
    #[global_allocator]
    static ALLOC: wee_alloc::WeeAlloc =
      wee_alloc::WeeAlloc::INIT;
    #[wasm_bindgen]
    pub fn is_palindrome(input: &str) -> bool {
        let s = input.to_string().to_lowercase();
        s == s.chars().rev().collect::<String>()
    }
    ```

We import wasm_bindgen and then enable wee_alloc, which does a much smaller memory allocation.

We go on to define the is_palindrome function, which takes &str as input and returns bool. Inside this function, we check whether the given string is a palindrome or not.

> **Note**
> Find out more about the difference between &str and String at https://users.rust-lang.org/t/whats-the-difference-between-string-and-str/10177/9.

4. Now, remove all the lines from js/index.js and replace them with the following content:

```
const rust = import('../pkg/index.js');
rust.then(module => {
    console.log(module.is_palindrome('tattarrattat'));
  // returns true
});
```

> **Note**
> We are importing from ../pkg/index.js here. The wasm-pack command will generate the binding file and wasm file inside the pkg folder.

5. Next, build the application with the following command:

```
$ npm run build
// comments, logs are elided
    Asset      Size    Chunks      Chunk Names
    0.js    9.84 KiB       0    [emitted]
    0fd5cbc32a547ac3295c.module.wasm    115 KiB        0
      [emitted] [immutable]
    index.html   179 bytes            [emitted]
    index.js    901 KiB    index    [emitted]
      index
```

You can run the application with the npm run start command. This command opens the browser and loads the application.

6.   Now, open the developer tools and check the logs in the console.

The WebAssembly module generated by the Rust compiler is not completely optimized. We can optimize the WebAssembly modules further. In the JavaScript world, every byte matters.

7.   Now, open `Cargo.toml` and add the following content:

```
[profile.dev]
opt-level = 'z'
lto = true
debug = false
```

Also, remove the `[profile.release]` section completely. The `[profile.dev]` section instructs the compiler on how to profile the code generated in the dev build. The `[profile.release]` section is used only for the release build.

We instruct the compiler to use `opt-level = z` for generating the code. The `opt-level` setting is similar to the LLVM compiler's `-O1/2/3/....`.

The valid options of the `opt-level` setting are as follows:

- `0` – no optimizations; also turns on `cfg(debug_assertions)`
- `1` – basic optimizations
- `2` – some optimizations
- `3` – all optimizations
- `s` – optimize for binary size
- `z` – optimize for binary size, but also turn off loop vectorization

LLVM supports link-time optimizations to better optimize code by using whole program analysis. But link-time optimization comes at the cost of longer linking time. We can enable LLVMs link-time optimization using the lto option.

The lto supports the following options:

- `false` – performs "thin local LTO". This means the link-time optimizations are done only on the local crate. Note: there will not be any link-time optimizations when the number of Codegen units is 1 or `opt-level` is 0.
- `true` or "fat" – performs "fat" LTO. This means the link-time optimizations are done across all crates in the dependency graph.

- `thin` – performs "thin" LTO. This is a faster version of "fat", that optimizes at a faster rate.

- `off` – disables LTO.

8. Next, run `npm run build`:

```
$ npm run build
// comments, logs are elided
   Asset    Size    Chunks    Chunk Names
   0.js  9.84 KiB      0    [emitted]
   b5e867dd3d25627d7122.module.wasm  50.8 KiB          0
     [emitted] [immutable]
   index.js    901 KiB    index    [emitted]
     index
```

The generated WebAssembly binary is 50.8 KB. The generated binary is ~44% smaller in size. That is a huge win for us. We can further optimize the binary using Binaryen's `wasm-opt` tool:

```
$ /path/to/build/directory/of/binaryen/wasm-opt -Oz
  b5e867dd3d25627d7122.module.wasm -o opt-gen.wasm
$ l
-rw-r--r--    1 sendilkumar  staff    45K May  8 17:43
  opt-gen.wasm
```

It reduces another 5 KB. We have used the `-Oz` pass, but we can pass in other passes to further optimize the generated binary.

We have seen how to minimize the WebAssembly module using Rust. Next, we will analyze the memory model in the WebAssembly module.

Analyzing the memory model in the WebAssembly module

Inside the JavaScript engine, WebAssembly and JavaScript run at different locations. Crossing the boundaries between JavaScript and WebAssembly will always have a cost attached to it. The browser vendors implemented cool hacks and workarounds to reduce this cost, but when your applications cross this boundary, this boundary crossing will often soon become a major performance bottleneck for your application. It is very important to design WebAssembly applications in a way that reduces boundary crossing. But once the application grows, it becomes difficult to manage this boundary crossing. To prevent boundary crossing, WebAssembly modules come with the memory module.

The memory section in the WebAssembly module is a vector of linear memories.

A linear memory model is a memory-addressing technique in which the memory is organized in a single contagious address space. It is also known as the Flat memory model.

The linear memory model makes it easier to understand, program, and represent the memory. But it also has huge disadvantages, such as high execution time for rearranging elements in memory and wasting a lot of memory area.

Here, the memory represents a vector of raw bytes containing uninterpreted data. WebAssembly uses resizable array buffers to hold the raw bytes of memory. It is important to note that this memory that is created is accessible and mutable from both JavaScript and WebAssembly.

Sharing memory between JavaScript and WebAssembly using Rust

We have already seen how to share the memory between JavaScript and WebAssembly. Let's share memory using Rust in this example:

1. Create a new Rust project using Cargo:

    ```
    $ cargo new --lib memory_world
    ```

2. Open the project in your favorite editor and replace src/lib.rs with the following content:

    ```
    #![no_std]

    use core::panic::PanicInfo;
    use core::slice::from_raw_parts_mut;
    ```

```rust
#[no_mangle]
fn memory_to_js() {
    let obj: &mut [u8];

    unsafe {
        obj = from_raw_parts_mut::<u8>(0 as *mut u8, 1);
    }

    obj[0] = 13;
}

#[panic_handler]
fn panic(_info: &PanicInfo) -> !{
    loop{}
}
```

The Rust file starts with #![no_std]. This instructs the compiler not to include the Rust Standard Library while generating the WebAssembly module. This will reduce the binary size a lot. Next, we define a function called memory_to_js. This function creates an obj in memory and shares it with JavaScript. In the function definition, we create a slice of u32 called obj. Next, we assign some raw memory to obj. Here, we are dealing with raw memory. Hence, we wrap the code inside an unsafe block. The memory object is global and it is mutable by both JavaScript and WebAssembly. Hence, we use from_raw_parts_mut to instantiate the object. Finally, we assign a value to the first element in the shared array buffer.

3. Create an index.html file and add the following content:

```html
<script>
    ( async() => {
        const bytes = await fetch("target/wasm32-
        unknown-unknown/debug/memory_world.wasm");
        const response = await bytes.arrayBuffer();
        const result = await
            WebAssembly.instantiate(response, {});
        result.exports.memory_to_js();
        const memObj = new
            UInt8Array(result.exports.memory.buffer, 0)
```

```
            .slice(0, 1);
          console.log(memObj[0]); // 13
    })();
</script>
```

We create an anonymous asynchronous JavaScript function that will be invoked as soon as the script is loaded. We fetch the WebAssembly binary inside the anonymous function. Next, we create `ArrayBuffer` and instantiate the module to the `result` object. We then call the `memory_to_js` method in the WebAssembly module (note the `exports` keyword, since the function is exported from the WebAssembly module). This instantiates the memory and assigns the first element in the shared array buffer to `13`:

```
const memObj = new
  UInt8Array(result.exports.memory.buffer, 0)
    .slice(0, 1);
console.log(memObj[0]); // 13
```

Next, we call the memory object that is exported from WebAssembly using `result.export.memory.buffer` and convert it into `UInt8Array` using a new `UInt8Array()`. Next, we extract the first element using `slice(0,1)`. This way, we can pass and retrieve values between JavaScript and WebAssembly without any overhead. The memory is accessed via `load` and `store` binary instructions. The `load` operation copies data from the main memory to register. The `store` operation copies data from the main memory. These binary instructions are accessed with the offset and the alignment. The alignment is in base-2 logarithmic representation. The memory address should be a multiple of four. This is called alignment restriction. This alignment restriction makes the hardware much faster.

> **Note**
>
> It is important to note that WebAssembly currently provides only a 32-bit address range. In the future, WebAssembly might provide a 64-bit address range.

We have seen how to share memory between JavaScript and WebAssembly by creating the memory in Rust. Next, we will create a memory object on the JavaScript side and use it inside the Rust application.

Creating a memory object in JavaScript to use in the Rust application

Unlike JavaScript, Rust is not dynamically typed. The memory created in JavaScript has no way to tell WebAssembly (or the Rust code) what to allocate and when to free them. We need to explicitly inform WebAssembly how to allocate the memory and, most importantly, when and how to free them (to avoid any leaks).

We use the `WebAssembly.memory()` constructor to create the memory in JavaScript. The memory constructor takes in an object to set the defaults. The object has the following options:

- `initial` – the initial size of the memory
- `maximum` – the maximum size of the memory (optional)
- `shared` – to denote whether to use the shared memory

The units for `initial` and `maximum` are WebAssembly pages, where a page refers to 64 KB.

We change the HTML file as follows:

```
<script>
    ( async() => {
        const memory = new WebAssembly.Memory({initial: 10,
            maximum:100}); // -> 1
        const bytes = await fetch("target/wasm32-unknown-
            unknown/debug/memory_world.wasm");
        const response = await bytes.arrayBuffer();
        const instance = await
            WebAssembly.instantiate(response,
            { js: { mem: memory } }); // ->2
        const s = new Set([1, 2, 3]);
        let jsArr = Uint8Array.from(s); // -> 3
        const len = jsArr.length;
        let wasmArrPtr = instance.exports.malloc(length);
            // -> 4
        let wasmArr = new
            Uint8Array(instance.exports.memory.buffer,
            wasmArrPtr, len); // -> 5
```

```
        wasmArr.set(jsArr); // -> 6
        const sum = instance.exports.accumulate
          (wasmArrPtr, len); // -> 7
        console.log(sum);
    })();
</script>
```

In `// -> 1`, the memory is initialized with the `WebAssembly.Memory()` constructor. We passed in the initial and maximum size of the memory, that is, 640 KB and 6.4 MB, respectively.

In `// -> 2`, we're instantiating the WebAssembly module along with the memory object.

In `// -> 3`, we then create `typedArray` (`UInt8Array`) with values 1, 2, and 3.

In `// -> 4`, we see how, as WebAssembly modules do not have any clue about the objects that are created out of the memory, the memory needs to be allocated. We have to manually write the allocation and freeing of memory in WebAssembly. In this step, we send the length of the array and allocate that memory. This gives us a pointer to the location of the memory.

In `// -> 5`, we create a new `typedArray` with the buffer (total available memory), the memory offset (`wasmAttrPtr`), and the length of the memory.

In `// -> 6`, we set the locally created `typedArray` (in *step 3*) to `typedArray` created in *step 5*.

In `//-> 7`, finally, we send the pointer to the memory and the length to the WebAssembly module, where we fetch the value from the memory by using the pointer to the memory and the length.

On the Rust side, replace the contents of `src/lib.rs` with the following:

```rust
use std::alloc::{alloc, dealloc, Layout};
use std::mem;

#[no_mangle]
fn accumulate(data: *mut u8, len: usize) -> i32 {
    let y = unsafe { std::slice::from_raw_parts(data as
      *const u8, len) };
    let mut sum = 0;
    for i in 0..len {
        sum = sum + y[i];
```

```
        }
        sum as i32
}


#[no_mangle]
fn malloc(size: usize) -> *mut u8 {
    let align = std::mem::align_of::<usize>();
    if let Ok(layout) = Layout::from_size_align(size,
      align) {
        unsafe {
            if layout.size() > 0 {
                let ptr = alloc(layout);
                if !ptr.is_null() {
                    return ptr
                }
            } else {
                return align as *mut u8
            }
        }
    }
    std::process::abort
}
```

We imported alloc, dealloc, and Layout from std::alloc and std::mem to play with the raw memory. The first function, accumulate, takes in data, which is the pointer where the data starts, and len, the length of the memory to read. First, we create a slice from the raw memory using std::slice::from_raw_parts by passing the pointer, data, and length, len. Note that this is an unsafe operation. Next, we run through the items in the array and add all the elements. Finally, we return the value as i32.

The malloc function is used to custom-allocate the memory since the WebAssembly module has no clue about the type of information sent and how to read/understand it. malloc helps us to allocate the memory as required without any panic.

Run the previous code using python -m http.server and load the web page in a browser to see the results in the developer tools.

Analyzing the WebAssembly module with Twiggy

The Rust-to-WebAssembly binaries are more likely to create a bloated binary. Proper care should be taken when creating WebAssembly binaries. The trade-off between the level of optimization, the time to compile, and various other factors should be considered while producing binaries. But most of the preceding work is done by the compiler by default. Either Emscripten or the `rustc` compiler ensures the elimination of dead code along with various options on the optimization level (`-O0` to z). We can choose the one that fits us.

Twiggy is a code size profiler. It uses the call graph to determine the origins of a function and provides meta-information about the function. The meta-information includes the size of each function in binary and its cost. Twiggy provides an overview of what is in the binary. With that information, we can optimize the binary further Let's install and use Twiggy to optimize the binary:.

1. Install Twiggy by running the following command:

    ```
    $ cargo install twiggy
    ```

2. Once installed, the `twiggy` command will be available in the command line, which we can check with the following command:

    ```
    $ twiggy
    twiggy-opt 0.6.0

    ...

    Use `twiggy` to make your binaries slim!

    USAGE:
        twiggy <SUBCOMMAND>

    FLAGS:
        -h, --help Prints help information
        -V, --version Prints version information

    SUBCOMMANDS:
        diff          Diff the old and new versions of a
                      binary to see what sizes changed.
        dominators    Compute and display the dominator
    ```

	tree for a binary's call graph.
garbage	Find and display code and data that is not transitively referenced by any exports or public functions.
help	Prints this message or the help of the given subcommand(s)
monos	List the generic function monomorphizations that are contributing to code bloat.
paths	Find and display the call paths to a function in the given binary's call graph.
top	List the top code size offenders in a binary.

3. Create a folder to test-drive Twiggy:

```
$ mkdir twiggy-world
```

4. Create a file called add.wat and add the following content:

```
$ touch add.wat
(module
    (func $add (param $lhs i32) (param $rhs i32)
      (result i32)
        get_local $lhs
        get_local $rhs
        i32.add)
    (export "add" (func $add))
)
```

5. Once you have defined the WebAssembly text format, compile it to the WebAssembly module using wabt:

```
$ /path/to/build/directory/of/wabt/wat2wasm add.wat
```

6. The preceding command generates an add.wasm file. To get the call paths in the binary, run Twiggy with the paths option:

```
$ twiggy paths add.wasm
Shallow Bytes | Shallow % | Retaining Paths
──────────────┼───────────┼─────────────────
            9 │   21.95%  │ code[0]
              │           │   ⌐ export "add"
            8 │   19.51%  │ wasm magic bytes
            6 │   14.63%  │ type[0]: (i32, i32) -> i32
              │           │   ⌐ code[0]
              │           │     ⌐ export "add"
            6 │   14.63%  │ export "add"
            6 │   14.63%  │ code section headers
            3 │    7.32%  │ type section headers
            3 │    7.32%  │ export section headers
```

The `twiggy paths` command shows the call path for the functions, the number of bytes they occupy inside the binary, and their percentage. The actual added code is 9 bytes and it occupies 21.95% of the total binary size.

Let's explore various subcommands in Twiggy:

- top
- monos
- garbage

top

The `twiggy top` command will list the code size of each block. It lists, in descending order, the size of the function, the percentage of the size in the end binary and the block section:

```
$ twiggy top add.wasm
Shallow Bytes | Shallow % | Item
──────────────┼───────────┼──────────────────
            9 │   21.95%  │ code[0]
            8 │   19.51%  │ wasm magic bytes
            6 │   14.63%  │ type[0]: (i32, i32) -> i32
```

```
        6 |      14.63% | export "add"
        6 |      14.63% | code section headers
        3 |       7.32% | type section headers
        3 |       7.32% | export section headers
       41 |     100.00% | Σ [7 Total Rows]
The usage of the twiggy top is as follows
USAGE: twiggy top <input> -n <max_items> -o
  <output_destination> --format <output_format> --mode
  <parse_mode>
```

List the top n details using -n followed by the number of entries to show:

```
$ twiggy top add.wasm -n 3
Shallow Bytes | Shallow % | Item
──────────────┼───────────┼──────────────────────────────
        9 |      21.95% | code[0]
        8 |      19.51% | wasm magic bytes
        6 |      14.63% | type[0]: (i32, i32) -> i32
       18 |      43.90% | ... and 4 more.
       41 |     100.00% | Σ [7 Total Rows]
```

Similarly, we can format the output to JSON format using the --format flag:

```
$ twiggy top add.wasm -n 3 --format json
[{"name":"code[0]","shallow_size":9,"shallow_size_percent":
21.951219512195124},{"name":"wasm magic
bytes","shallow_size":8,"shallow_size_percent":19.512195121
95122},{"name":"type[0]: (i32, i32) ->
i32","shallow_size":6,"shallow_size_percent":14.63414634146
3413}]
```

The top command is extremely useful when you want to track down the biggest code blocks and then optimize them separately.

monos

In the JavaScript world, monomorphization increases performance. But it also bloats the code size (for example, in generics). Since we have to create the implementation of a generic function dynamically for every type, we have to be very careful when using generics and monomorphic code.

Twiggy has a subcommand called `monos` that will list the code bloating due to monomorphization:

```
$ twiggy monos pkg/index_bg.wasm
Apprx. Bloat Bytes │ Apprx. Bloat % │ Bytes │ %        │
Monomorphizations

─────────────────────┼─────────────────┼────────┼──────────┼────────────

                 4 │          0.01% │    32 │  0.06% │
core::ptr::drop_in_place
                   │                 │    28 │  0.05% │
core::ptr::drop_in_place::h9684ba572bb4c2f9
                   │                 │     4 │  0.01% │
core::ptr::drop_in_place::h00c08aab80423b88
                 0 │          0.00% │  5437 │ 10.44% │
dlmalloc::dlmalloc::Dlmalloc::malloc
                   │                 │  5437 │ 10.44% │
dlmalloc::dlmalloc::Dlmalloc::malloc::hb0329e71e24f7e2f
                 0 │          0.00% │  1810 │  3.48% │ <char
as core::fmt::Debug>::fmt
                   │                 │  1810 │  3.48% │
<char as core::fmt::Debug>::fmt::h5472f29c33f4c4c9
                 0 │          0.00% │  1126 │  2.16% │
dlmalloc::dlmalloc::Dlmalloc::free
                   │                 │  1126 │  2.16% │
dlmalloc::dlmalloc::Dlmalloc::free::h7ab57ecacfa2b1c3
                 0 │          0.00% │  1123 │  2.16% │
core::str::slice_error_fail
                   │                 │  1123 │  2.16% │
core::str::slice_error_fail::h26278b2259fb6582
                 0 │          0.00% │   921 │  1.77% │
core::fmt::Formatter::pad
                   │                 │   921 │  1.77% │
core::fmt::Formatter::pad::hb011277a1901f9f7
```

```
                0 |             0.00% |   833 |   1.60% |
dlmalloc::dlmalloc::Dlmalloc::dispose_chunk
                  |                   |   833 |   1.60% |
dlmalloc::dlmalloc::Dlmalloc::dispose_chunk::he00c681454a3c3b7
                0 |             0.00% |   787 |   1.51% |
core::fmt::write
                  |                   |   787 |   1.51% |
core::fmt::write::hb395f946a5ce2cab
                0 |             0.00% |   754 |   1.45% |
core::fmt::Formatter::pad_integral
                  |                   |   754 |   1.45% |
core::fmt::Formatter::pad_integral::h05ee6133195a52bc
                0 |             0.00% |   459 |   0.88% |
alloc::string::String::push
                  |                   |   459 |   0.88% |
alloc::string::String::push::he03a5b89b77597a1
                0 |             0.00% |  4276 |   8.21% | ... and
64 more.
                4 |             0.01% | 17558 |  33.73% | Σ [85
Total Rows]
....
```

We are using the index_bg.wasm example from the *Minimizing the WebAssembly modules* section of this chapter.

monos is extremely useful for us to understand the occurrence of any bloating caused by generic parameters, which can then be changed to a simpler function with generics.

garbage

At times, it is important to find code that is not used anymore but is kept in the final binary due to some other reasons. These functions are referenced somewhere but not used anywhere and the compiler will not know when and where to remove them.

We can use Twiggy's `garbage` command to list all the code and data that is not transitively referenced:

```
$ twiggy garbage add.wasm
 Bytes | Size % | Garbage Item
-------+--------+-------------------------------------------
   109 |  0.21% | custom section 'producers'
   109 |  0.21% | Σ [1 Total Rows]
 27818 | 53.44% | 1 potential false-positive data segments
```

WebAssembly modules consist of a data section. But sometimes, we might not use the data straight away in the WebAssembly module but in some other places where it is imported. As you can see here, Twiggy's `garbage` subcommand shows those potentially false values.

Summary

In this chapter, we have seen how to optimize the WebAssembly binary using Rust, how to map memory between JavaScript and Rust, and finally, how to analyze a WebAssembly module using Twiggy.

The WebAssembly ecosystem is still in its early days and it promises better performance. The WebAssembly binary addresses a few gaps in the JavaScript ecosystem, such as size-efficient compact binaries, enabling streaming compilation, and properly typed binaries. These features make WebAssembly smaller and faster. Rust, on the other hand, provides first-in-class support for generating a WebAssembly module and `wasm-bindgen` is the best tool available that makes it easier to transfer complex objects in Rust and WebAssembly.

I hope that you now understand the basics of WebAssembly and how Rust makes it easier to generate WebAssembly modules. I can't wait to see what you will be shipping with Rust and WebAssembly.

Index

Symbols

--fold-exprs 66
-f option 66
--profiling flag
  reference link 31
&str and String
  reference link 189

A

abstract syntax trees (ASTs) 37, 181
asm.js
  about 12
  generating, with Emscripten 15, 16

B

backend 5, 6
Binaryen
  about 101
  installing 102
  installing, on Linux/MacOSx 103
  installing, on Windows 103
  tools 104
  using 102
binding file 19

C

Cargo
  about 118
  Rust, converting into
    WebAssembly via 123-126
Clang 12
classes
  sharing, from JavaScript
    with Rust 164-169
  sharing, from Rust with
    JavaScript 160-164
Closure Compiler
  about 18, 32, 33
  reference link 34
closures
  calling, via WebAssembly 173-176
CMake
  download link 55
CMake projects, in Visual Studio
  reference link 103
code splitting
  reference link 39
compilation 4
compile phase 39

compiler
    about 4, 5
    efficiency 6, 7
compiler, components
    backend 6
    frontend 5
    optimizer 6
C source code
    WebAssembly module (WASM),
        converting into 67

D

Debugging Information Entry (DIE) 8
decode phase 38
DWARF
    about 8
    reference link 8

E

emrun
    reference link 22
Emscripten
    installing, with Emscripten
        SDK (emsdk) 13, 14
    used, for generating asm.js 15, 16
    used, for running Hello World
        in browser 20-22
    used, for running Hello
        World in Node 17-20
Emscripten-compiled pages
    reference link 22
Emscripten compiler frontend (emcc)
    about 12
    reference link 34

Emscripten SDK (emsdk)
    about 12
    options, exploring 22, 23
    tools and SDK, listing 24, 25
    tools and SDK, managing 25, 26
    used, for installing Emscripten 13, 14
Emscripten SDK (emsdk), options
    number of cores to build 26
    tools and SDK, activating 27
    tools and SDK, uninstalling 27
    type of build 26, 27
execute phase 39
export keyword 44
exports section 84-87

F

Fibonacci function
    building, in WebAssembly
        text format 45-47
frontend 5
func keyword 42
func_type 43

G

garbage collection 38
garbage command 203
globals 89-95

H

Hello World
    running, with Emscripten
        in browser 20-22
    running, with Emscripten
        in Node 17-20

high-level intermediate
      representation (HIR) 122

I

Immediately Invoked Function
      Execution (IIFE) 125
imports section 87, 88
Instruction Set Architecture (ISA) 6, 36
Intermediate Representation (IR) 5, 101
interpreted languages 5
interpret phase 37

J

JavaScript
   disadvantages 3
   reference link 39
   Rust, using to share classes
      from 164-169
   used, for sharing classes
      from Rust 160-164
JavaScript API
   calling, via WebAssembly 169-173
JavaScript engine
   JavaScript execution 36-38
   WebAssembly execution 38, 39
JavaScript function
   importing, into Rust 177-180
JSON
   WebAssembly text (WAST),
      converting into 73-75
just-in-time (JIT) 37

L

large codebases, optimizing
      with Emscripten
   reference link 34
linear memory model 192
LLVM core 7
LLVM intermediate representation
      (LLVM IR) 7, 12, 122
loop keyword 48
Low-Level Virtual Machine (LLVM)
   advantages 7
   exploring 7
   URL 10
   working 8, 9
lto setting
   options 190

M

machine code 6
Macros 120
memory model
   analyzing, in WebAssembly
      module 192-197
memory section 96-100
middle intermediate representation
      (MIR) 122
monos command 202, 203

O

objdump
   reference link 58
optimization options
   reference link 34

optimizations
  about 28
  Closure Compiler 32, 33
  levels 28
  options 28
  options, exploring 29-32
optimize phase 37
optimizer 5, 6
opt-level setting
  options 190

P

param keyword 43
Parcel
  used, for bundling WebAssembly
    modules 145-149
parse phase 37, 38
profilers 37

R

Rust
  converting, into WebAssembly
    via Cargo 123-126
  converting, into WebAssembly
    via rustc 120-122
  converting, into WebAssembly
    via wasm-bindgen 129-135
  installing 118-120
  JavaScript function, importing
    into 177-180
  JavaScript, using to share
    classes from 160-164
  used, for sharing classes from
    JavaScript 164-169
Rust compiler (rustc) 118

S

s-expressions
  reference link 50
simple.c 69- 72
simple.h 68, 69
start function 95, 96

T

tools
  reference link 27
top command 200, 201
Twiggy
  used, for analyzing WebAssembly
    module 198-200
Twiggy subcommands
  garbage command 203
  monos command 202, 203
  top command 200, 201
TypeScript 38
type_use 43

U

unsigned 32-bit (u32) 42

W

wasm2js tool 111-113
wasm-as tool 105-107
wasm-bindgen
  installing 127-129
wasm-dis tool 107-109
wasm-interp 81, 82
wasm-objdump 75-77
wasm-opt tool 109-111

wasm-pack
  about 150
  need for 150
  used, for packing 153-156
  used, for publishing 153-156
  using 150-153
wasm-pack publish command, options
  --access 156
  --target 156
wasm-strip 78, 79
wasm-validate 79
watchers 37
web API
  calling, via WebAssembly 180-184
WebAssembly
  about 4
  closures, calling via 173-176
  in JavaScript engine 39
  JavaScript API, calling via 169-173
  rust, converting via Cargo 123-126
  rust, converting via rustc 120-122
  rust, converting via
    wasm-bindgen 129-135
  working 36
WebAssembly Binary Toolkit (WABT)
  installing 54, 55
  installing, on Linux 55, 56
  installing, on macOS 55, 56
  installing, on Windows 56-58
  tools 75
WebAssembly execution
  in JavaScript engine 38
webassemblyjs 140
WebAssembly module (WASM)
  about 35
  analyzing, with Twiggy 198-200
  bundling, with Parcel 145-149
  bundling, with webpack 138-145

converting, into C 67
converting, into WebAssembly
  text (WAST) 62-65
exports section 84-87
globals section 89-95
imports section 87, 88
memory model, analyzing 192-197
memory section 96-100
optimizing 188-191
sections 41
start function 95, 96
WebAssembly text (WAST),
  converting into 58-61
WebAssembly specification
defining 36
reference link 20
WebAssembly text (WAST)
converting, into JSON 73-75
converting, into WebAssembly
  module (WASM) 58-61
exploring 40-44
Fibonacci function, building 45-47
loop, creating 48, 49
reference link 50
WebAssembly module (WASM),
  converting into 62-65
webpack
used, for bundling WebAssembly
  modules 138-145
web servers
reference link 21

Other Books You May Enjoy

If you enjoyed this book, you may be interested in these other books by Packt:

Creative Projects for Rust Programmers

Carlo Milanesi

ISBN: 978-1-78934-622-0

- Access TOML, JSON, and XML files and SQLite, PostgreSQL, and Redis databases
- Develop a RESTful web service using JSON payloads
- Create a web application using HTML templates and JavaScript and a frontend web application or web game using WebAssembly
- Build desktop 2D games
- Develop an interpreter and a compiler for a programming language
- Create a machine language emulator
- Extend the Linux Kernel with loadable modules

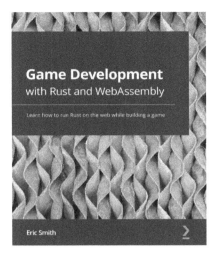

Game Development with Rust and WebAssembly

Eric Smith

ISBN: 978-1-80107-097-3

- Build and deploy a Rust application to the web using WebAssembly
- Use wasm-bindgen and the Canvas API to draw real-time graphics
- Write a game loop and take keyboard input for dynamic action
- Explore collision detection and create a dynamic character that can jump on and off platforms and fall down holes
- Manage animations using state machines
- Generate levels procedurally for an endless runner
- Load and display sprites and sprite sheets for animations
- Test, refactor, and keep your code clean and maintainable

Packt is searching for authors like you

If you're interested in becoming an author for Packt, please visit `authors.packtpub.com` and apply today. We have worked with thousands of developers and tech professionals, just like you, to help them share their insight with the global tech community. You can make a general application, apply for a specific hot topic that we are recruiting an author for, or submit your own idea.

Share Your Thoughts

Now you've finished *Practical WebAssembly*, we'd love to hear your thoughts! Scan the QR code below to go straight to the Amazon review page for this book and share your feedback or leave a review on the site that you purchased it from.

`https://packt.link/r/1838828001`

Your review is important to us and the tech community and will help us make sure we're delivering excellent quality content.

www.ingramcontent.com/pod-product-compliance
Lightning Source LLC
Chambersburg PA
CBHW060550060326
40690CB00017B/3661